This series aims to report new developments in mathematical research and teaching quickly, informally and at a high level. The type of material considered for publication includes:

1. Preliminary drafts of original papers and monographs
2. Lectures on a new field, or presenting a new angle on a classical field
3. Seminar work-outs
4. Reports of meetings

Texts which are out of print but still in demand may also be considered if they fall within these categories.

The timeliness of a manuscript is more important than its form, which may be unfinished or tentative. Thus, in some instances, proofs may be merely outlined and results presented which have been or will later be published elsewhere.

Publication of *Lecture Notes* is intended as a service to the international mathematical community, in that a commercial publisher, Springer-Verlag, can offer a wider distribution to documents which would otherwise have a restricted readership. Once published and copyrighted, they can be documented in the scientific literature.

Manuscripts

Manuscripts are reproduced by a photographic process; they must therefore be typed with extreme care. Symbols not on the typewriter should be inserted by hand in indelible black ink. Corrections to the typescript should be made by sticking the amended text over the old one, or by obliterating errors with white correcting fluid. Should the text, or any part of it, have to be retyped, the author will be reimbursed upon publication of the volume. Authors receive 75 free copies.

The typescript is reduced slightly in size during reproduction; best results will not be obtained unless the text on any one page is kept within the overall limit of 18 x 26.5 cm (7 x 10 ½ inches). The publishers will be pleased to supply on request special stationery with the typing area outlined.

Manuscripts in English, German or French should be sent to Prof. Dr. A. Dold, Mathematisches Institut der Universität Heidelberg, Tiergartenstraße or Prof. Dr. B. Eckmann, Eidgenössische Technische Hochschule, Zürich.

Die *„Lecture Notes"* sollen rasch und informell, aber auf hohem Niveau, über neue Entwicklungen der mathematischen Forschung und Lehre berichten. Zur Veröffentlichung kommen:

1. Vorläufige Fassungen von Originalarbeiten und Monographien.
2. Spezielle Vorlesungen über ein neues Gebiet oder ein klassisches Gebiet in neuer Betrachtungsweise.
3. Seminarausarbeitungen.
4. Vorträge von Tagungen.

Ferner kommen auch ältere vergriffene spezielle Vorlesungen, Seminare und Berichte in Frage, wenn nach ihnen eine anhaltende Nachfrage besteht.

Die Beiträge dürfen im Interesse einer größeren Aktualität durchaus den Charakter des Unfertigen und Vorläufigen haben. Sie brauchen Beweise unter Umständen nur zu skizzieren und dürfen auch Ergebnisse enthalten, die in ähnlicher Form schon erschienen sind oder später erscheinen sollen.

Die Herausgabe der *„Lecture Notes"* Serie durch den Springer-Verlag stellt eine Dienstleistung an die mathematischen Institute dar, indem der Springer-Verlag für ausreichende Lagerhaltung sorgt und einen großen internationalen Kreis von Interessenten erfassen kann. Durch Anzeigen in Fachzeitschriften, Aufnahme in Kataloge und durch Anmeldung zum Copyright sowie durch die Versendung von Besprechungsexemplaren wird eine lückenlose Dokumentation in den wissenschaftlichen Bibliotheken ermöglicht.

Lecture Notes in Mathematics

A collection of informal reports and seminars
Edited by A. Dold, Heidelberg and B. Eckmann, Zürich

104

George H. Pimbley, Jr.

University of California
Los Alamos Scientific Laboratory, Los Alamos, New Mexico

Eigenfunction Branches of Nonlinear Operators, and their Bifurcations

Springer-Verlag
Berlin · Heidelberg · New York 1969

Work performed under the auspices of the U. S. Atomic Energy Commission

Library of Congress Catalog Card Number 70-97958 Title No. 3710

TABLE OF CONTENTS

INTRODUCTION

The series of lectures on nonlinear operators covered by these lecture notes was given at the Battelle Memorial Institute Advanced Studies Center in Geneva, Switzerland during the period June 27 - August 5, 1968, at the invitation of Dr. Norman W. Bazley of the Battelle Research Center in Geneva. The material is taken from the results of approximately seven years of work on the part of the author at the Los Alamos Scientific Laboratory of the University of California, Los Alamos, New Mexico. Much of this material had previously been published in the open literature (see the Bibliography). This effort was generated by the need for a nonlinear theory observed in connection with actual problems in physics at Los Alamos.

In deriving nonlinear theory, abstract formulation is perhaps a desired end; but in the newer parts of the theory, as with secondary bifurcation in these notes, progress seems to be made more easily with concrete assumptions, as with our preoccupation with Hammerstein operators with oscillation kernels.

The entire lecture series had to do with the eigenvalue problem $\lambda x = T(x)$, where $T(x)$ is a bounded nonlinear operator. Other authors, with a view to applications in nonlinear differential equations with appropriate use of Sobolev spaces to render the operators bounded, have preferred to study eigenvalue problems of the form $(L_1+N_1)u = \lambda(L_2+N_2)u$, where L_1,L_2 are linear and N_1,N_2 are nonlinear. Such is the case with M. S. Berger [ref. 4]. In these notes we had the less ambitious goal of understanding nonlinear integral equations, whence we concentrated on the simpler form $\lambda x = T(x)$.

In the more abstract part of the notes, X is any Banach space with real scalars. When we take up Hammerstein operators, the space is the set C(0,1) of continuous real functions x(s) defined on the interval [0,1] with norm $\|x\| = \sup_{0\le s\le 1} |x(s)|$. When inner products are in evidence and when variational principles are used to study eigenvalues, it will be clear that the space $L_2(0,1)$ of functions square integrable on the interval [0,1] is in the background, and we are using the setwise embedding $C(0,1) \subset L_2(0,1)$.

The author wishes to thank Dr. Norman W. Bazley and Dr. Bruno Zwahlen of the Battelle Research Center, Geneva, for the opportunity to visit and present these lectures. He wishes to thank them also for many helpful suggestions. Thanks for good suggestions is also owed to Prof. Melvyn S. Berger, University of Minnesota; Prof. David W. Dean of Ohio State University, Prof. William G. Faris of Cornell University, Prof. Karl E. Gustafson of the University of Colorado, Prof. K. Kirchgässner of Friburg University, and Prof. Joachim Weidmann of the University of Munich. Appreciation is accorded also to Mrs. Margaret A. Gore of Los Alamos whose skill in typing a technical manuscript is amply demonstrated herein, and to Mrs. Barbara C. Powell for the various drawings.

The attention of the reader is directed to the summary of results in Section 10.

1. An Example.

So as to illustrate the type of problems considered in these notes, we present an eigenvalue problem for a nonlinear operator which can be attacked by elementary methods. Namely, we solve the following integral equation

$$\lambda\varphi(s) = \frac{2}{\pi}\int_0^\pi [a \sin s \sin t + b \sin 2s \sin 2t]\,[\varphi(t) + \varphi^3(t)]dt \quad (1.1)$$

which has a second-rank kernel. We suppose that $0 < b < a$. Because of the form of the kernel, any solution of eq. (1.1) is necessarily of the form $\varphi(s) = A \sin s + B \sin 2s$ with undetermined constants A,B (which will turn out to be functions of the real parameter λ). Substituting in eq. (1.1), we have

$$\lambda[A \sin s + B \sin 2s] = \frac{2}{\pi}\int_0^\pi [a \sin s \sin t + b \sin 2s \sin 2t]$$

$$\cdot\,[(A \sin t + B \sin 2t) + (A \sin t + B \sin 2t)^3]dt$$

$$= \frac{2}{\pi} a \sin s\left[A\int_0^\pi \sin^2 t\,dt + A^3\int_0^\pi \sin^4 t\,dt + 3AB^2\int_0^\pi \sin^2 t \sin^2 2t\,dt\right]$$

$$+ \frac{2}{\pi} b \sin 2s\left[B\int_0^\pi \sin^2 2t\,dt + 3A^2B\int_0^\pi \sin^2 2t \sin^2 t\,dt + B^3\int_0^\pi \sin^4 2t\,dt\right]$$

$$= \frac{2}{\pi} a \sin s\left[\frac{\pi}{2} A + \frac{3\pi}{8} A^3 + \frac{3\pi}{4} AB^2\right]$$

$$+ \frac{2}{\pi} b \sin 2s\left[\frac{\pi}{2} B + \frac{3\pi}{4} A^2B + \frac{3\pi}{8} B^3\right],$$

where use has been made of the following values of integrals:

$$\int_0^\pi \sin t \sin 2t \, dt = \int_0^\pi \sin^3 t \sin 2t \, dt = \int_0^\pi \sin t \sin^3 2t \, dt = 0.$$

Equating coefficients of sin s and sin 2s, we obtain a pair of nonlinear simultaneous algebraic equations:

$$\begin{aligned} \lambda A &= aA + \tfrac{3}{4} aA^3 + \tfrac{3}{2} aAB^2 \\ \lambda B &= bB + \tfrac{3}{2} bA^2B + \tfrac{3}{4} bB^3. \end{aligned} \tag{1.2}$$

There are four kinds of solutions of equations (1.2):

1) $A = B = 0$; this gives the trivial solution of eq. (1.1).

2) $A \neq 0$, $B = 0$; only the first equation is nontrivial. We cancel $A \neq 0$ to obtain

$$\lambda = a + \tfrac{3}{4} aA^2$$

whence

$$A = \pm \frac{2}{\sqrt{3}} \sqrt{\frac{\lambda}{a} - 1} \; .$$

The corresponding solution of eq. (1.1) is $\varphi_1(s,\lambda) = \pm \frac{2}{\sqrt{3}}\sqrt{\frac{\lambda}{a} - 1} \sin s$, defined and real for $\lambda \geq a$.

3) $A = 0$, $B \neq 0$; only the second equation is nontrivial. We cancel $B \neq 0$ to obtain

$$\lambda = b + \tfrac{3}{4} bB^2$$

whence

$$B = \pm \frac{2}{\sqrt{3}} \sqrt{\frac{\lambda}{b} - 1} \; .$$

The corresponding solution of eq. (1.1) is $\varphi_2(s,\lambda) = \pm \frac{2}{\sqrt{3}} \sqrt{\frac{\lambda}{b} - 1} \sin 2s$, defined and real for $\lambda \geq b$, where we recall that $b < a$.

4) $A \neq 0$, $B \neq 0$; here both A and B may be cancelled in eq. (1.2). We obtain the two ellipses:

$$\frac{3}{4}A^2 + \frac{3}{2}B^2 = \frac{\lambda}{a} - 1$$
$$\frac{3}{2}A^2 + \frac{3}{4}B^2 = \frac{\lambda}{b} - 1. \qquad (1.3)$$

Solutions of eq. (1.2) are given by intersections of these ellipses. Solving, we get

$$A^2 = \frac{4}{9}\left[\frac{2a-b}{ab}\lambda - 1\right], \qquad B^2 = \frac{4}{9}\left[\frac{2b-a}{ab}\lambda - 1\right],$$

so that we have the following solutions of eq. (1.1):

$$\varphi_3(s,\lambda) = \pm\frac{2}{3}\sqrt{\frac{2a-b}{ab}\lambda - 1}\ \sin s \pm \frac{2}{3}\sqrt{\frac{2b-a}{ab}\lambda - 1}\ \sin 2s. \qquad (1.4)$$

Clearly $2a-b > 0$ since we assumed that $b < a$. Hence the question of whether or not solutions of the form (1.4) can be real hinges upon whether or not $2b-a > 0$, or $\frac{b}{a} > \frac{1}{2}$. We have the following cases:

Case I: $\frac{b}{a} \leq \frac{1}{2}$; $\varphi_3(s,\lambda)$ is real for no real λ.

Case II: $\frac{b}{a} > \frac{1}{2}$; $\varphi_3(s,\lambda)$ is real for $\lambda > \max\left(\frac{ab}{2a-b}, \frac{ab}{2b-a}\right)$.

Since $a > b$, this means that $\varphi_3(s,\lambda)$ is real when $\lambda > \frac{ab}{2b-a}$.

Under case I above, i.e., when $\frac{b}{a} \leq \frac{1}{2}$, the only real solutions of eq. (1.1) are the trivial solution $\varphi(s,\lambda) \equiv 0$, and the two main branches:

$$\varphi_1(s,\lambda) = \pm\frac{2}{\sqrt{3}}\sqrt{\frac{\lambda}{a} - 1}\ \sin s$$
$$\varphi_2(s,\lambda) = \pm\frac{2}{\sqrt{3}}\sqrt{\frac{\lambda}{b} - 1}\ \sin 2s.$$

The solutions φ_1 and φ_2 branch away from the trivial solution $\varphi \equiv 0$ at the eigenvalues a,b of the linearization of eq. (1.1) at the origin:

$$\lambda h(s) = \frac{2}{\pi}\int_0^{\pi} [a \sin s \sin t + b \sin 2s \sin 2t]h(t)dt. \qquad (1.5)$$

We can represent this situation pictorially in two ways:

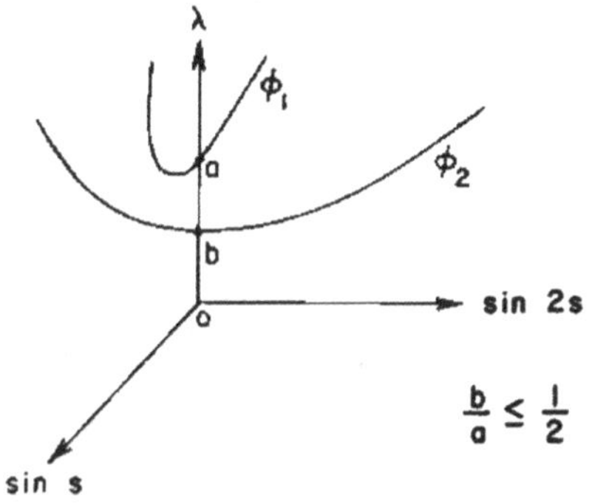

FIG. 1.1a.

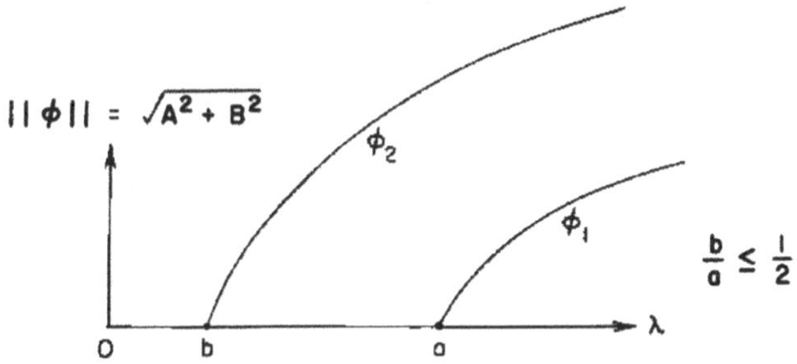

FIG. 1.1b.

In Fig. 1.1a we simply plot the two normal modes vs. the parameter λ, which serves well for eq. (1.1). In more general problems, as for example if we were to add to the rank of the kernel, a plot such as Fig. 1.1b must be used, where some norm of the solution is plotted vs. the parameter λ.

Under Case II above, i.e. when $\frac{b}{a} > \frac{1}{2}$, we again have the trivial solution $\varphi(s,\lambda) \equiv 0$, and the two main branches

$$\varphi_1(s,\lambda) = \pm \frac{2}{\sqrt{3}} \sqrt{\frac{\lambda}{a} - 1} \;\sin s$$

$$\varphi_2(s,\lambda) = \pm \frac{2}{\sqrt{3}} \sqrt{\frac{\lambda}{b} - 1} \;\sin 2s$$

which bifurcate from $\varphi \equiv 0$ at the primary bifurcation points, which are the eigenvalues a,b of linearized eq. (1.5). Moreover, for $\lambda > \frac{ab}{2b-a} > a$, a third type of solution branch appears, namely that in eq. (1.4). Note that as $\lambda \to \frac{ab}{2b-a}$, $\lambda > \frac{ab}{2b-a}$, the coefficients $\sqrt{\frac{2b-a}{ab}\lambda - 1} \to 0$ and $\sqrt{\frac{2a-b}{ab}\lambda - 1} \to \sqrt{\frac{3a-3b}{2b-a}}$. On the other hand note that $\sqrt{\frac{\lambda}{a} - 1} \to \sqrt{\frac{a-b}{2b-a}}$ as $\lambda \to \frac{ab}{2b-a}$. Thus as $\lambda \to \frac{ab}{2b-a}$, we see that $\varphi_3(s,\lambda) \to \varphi_3\left(s, \frac{ab}{2b-a}\right) = \varphi_1\left(s, \frac{ab}{2b-a}\right)$. Therefore at $\lambda = \frac{ab}{2b-a}$, the sub-branch (twig)

$$\varphi_3^+(s,\lambda) = \frac{2}{3} \sqrt{\frac{2a-b}{ab}\lambda - 1} \;\sin s \pm \frac{2}{3} \sqrt{\frac{2b-a}{ab}\lambda - 1} \;\sin 2s$$

joins the main branch, i.e., $\varphi_3^+\left(s, \frac{ab}{2b-a}\right) = \varphi_1^+\left(s, \frac{ab}{2b-a}\right)$ while the sub-branch (twig)

$$\varphi_3^-(s,\lambda) = -\frac{2}{3} \sqrt{\frac{2a-b}{ab}\lambda - 1} \;\sin s \pm \frac{2}{3} \sqrt{\frac{2b-a}{ab}\lambda - 1} \;\sin 2s$$

joins the negative part of the main branch, i.e., $\varphi_3^-\left(s, \frac{ab}{2b-a}\right) = \varphi_1^-\left(s, \frac{ab}{2b-a}\right)$.

We have here, under Case II, when $\frac{b}{a} > \frac{1}{2}$, the phenomena of "secondary bifurcation," or the forming of sub-branches or twigs which bifurcate from the main branches. The main branches bifurcate from the trivial solution at the eigenvalues of the linearization, eq. (1.5), while the twigs bifurcate from the main branches. We can represent the situation again in two ways:

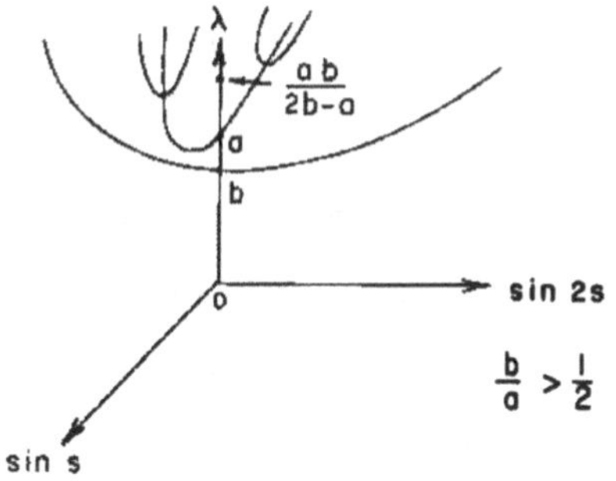

FIG. 1.2a.

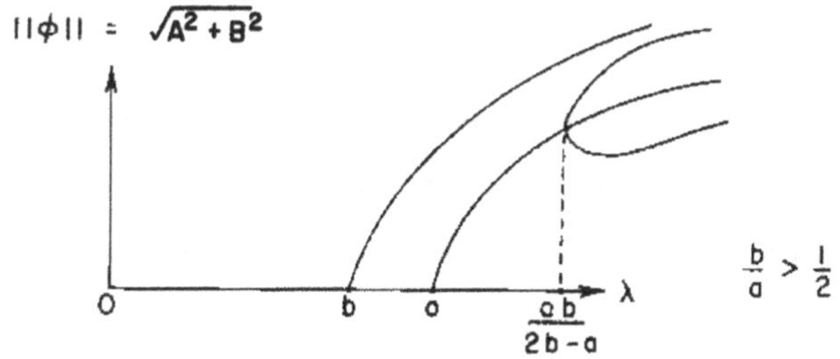

FIG. 1.2b.

Thus solutions of the nonlinear equation (1.1) exist as continuous loci in (λ, sin s, sin 2s) space. There are two main branches: $\varphi_1(s,\lambda)$ splits off from the trivial solution $\varphi \equiv 0$ at $\lambda = a$, and its two parts φ_1^+, φ_1^- differ only in sign; $\varphi_2(s,\lambda)$ joins the trivial solution at $\lambda = b$, and its two parts φ_2^+, φ_2^- differ only in sign. a and b on the λ axis are the primary bifurcation points for the main branches.

If $\frac{b}{a} > \frac{1}{2}$, i.e. Case II, two sub-branches or twigs split away from $\varphi_1(s,\lambda)$ at $\lambda = \frac{ab}{2b-a}$, which is known as a secondary bifurcation point.

The question of whether or not secondary bifurcation of the eigensolutions of eq. (1.1) takes place therefore hinges on whether we have $\frac{b}{a} > \frac{1}{2}$, or $\frac{b}{a} \leq \frac{1}{2}$. The condition $\frac{b}{a} \leq \frac{1}{2}$ in this simple problem is a "condition preventing secondary bifurcation." Much interest attaches generally to the question of whether we have secondary bifurcation of a given branch of eigensolutions, or of any branch of eigensolutions of a nonlinear eigenvalue problem, and to the derivation of conditions preventing or allowing secondary bifurcation. The occurrence of secondary bifurcation clearly has a marked effect on the matter of multiplicity of solutions, over considerable ranges of the real parameter λ, as this simple example shows.

The example of this section is such that the solutions can be completely worked out by elementary methods. In the next sections we present the methods of nonlinear functional analysis which must be employed to study bifurcations and solution branches in the general theory of nonlinear eigenvalue problems. There happens to be however much qualitative similarity between the structure of solutions of problem (1.1) of this section, and more general cases.

2. The Extension of Branches of Solutions for Nonlinear Equations in Banach Spaces.

In this section we consider general bounded continuously Fréchet-differentiable transformations $T(x)$ of a real Banach space X into itself: $x \in X$, $T(x) \in X$. We assume that $T(\theta) = \theta$, where θ is the null element. Let us suppose that the equation

$$\lambda x = T(x) + f, \tag{2.1}$$

where λ is a real parameter and $f \in X$ is a fixed element, has a solution $x_0 \in X$ for a value λ_0 of the parameter; i.e., suppose that $\lambda_0 x_0 = T(x_0) + f$. We pose the problem of finding a nearby solution $x_0 + h$ for a nearby value $\lambda = \lambda_0 + \delta$. Thus we solve the following equation for h,δ:

$$T(x_0+h) + f = (\lambda_0+\delta)(x_0+h). \tag{2.2}$$

Using the definition of the Fréchet derivative $T'(x_0)$ [ref. 15; p. 183], we can write eq. (2.2) in the form

$$T(x_0) + T'(x_0)h + R_1(x_0,h) + f = \lambda_0 x_0 + \lambda_0 h + \delta x_0 + \delta h$$

where

$$\frac{\|R_1(x_0,h)\|}{\|h\|} \to 0$$

as $\|h\| \to 0$. Using the assumption that λ_0, x_0 satisfy eq. (2.1) we have

$$[\lambda_0 I - T'(x_0)]h = -\delta x_0 - \delta h + R_1(x_0,h). \tag{2.3}$$

Since $T'(x_0)$ is a bounded linear transformation such that $T'(x_0)h \in X$ if $h \in X$, let us assume that $\lambda_0 \in \rho(T'(x_0))$; other complementary assumptions will be discussed in the next section. Thus $\lambda_0 I - T'(x_0)$ has a continuous inverse M. Then from eq. (2.3) we write

$$h = [\lambda_0 I - T'(x_0)]^{-1}\{-\delta x_0 - \delta h + R_1(x_0,h)\} = MF_\delta(h). \tag{2.4}$$

We now prove a preliminary result about $F_\delta(h)$ defined in eq. (2.4):

Lemma 2.1: The function $F_\delta(h) = -\delta x_0 - \delta h + R_1(x_0,h)$ satisfies a Lipschitz condition

$$\|F_\delta(h_1) - F_\delta(h_2)\| \le A(\delta,h_1,h_2)\|h_1-h_2\|,$$

with $A(\delta,h_1,h_2) > 0$, and $A(\delta,h_1,h_2) \to 0$ as $|\delta| \to 0$, $\|h_1\| \to 0$, $\|h_2\| \to 0$.

Proof: By definition of the Fréchet derivative,

$$R_1(x_0,h) = T(x_0+h) - T(x_0) - T'(x_0)h.$$

Hence

$$\begin{aligned} R_1(x_0,h_1) - R_1(x_0,h_2) &= T(x_0+h_1) - T(x_0+h_2) - T'(x_0)(h_1-h_2) \\ &= T(x_0+h_2+[h_1-h_2]) - T(x_0+h_2) - T'(x_0)(h_1-h_2) \\ &= T'(x_0+h_2)(h_1-h_2) + R_1(x_0+h_2,h_1-h_2) \\ &\qquad - T'(x_0)(h_1-h_2), \end{aligned}$$

so that

$$\|R_1(x_0,h_1)-R_1(x_0,h_2)\| \le \left\{\|T'(x_0+h_2)-T'(x_0)\| + \frac{\|R_1(x_0+h_2,h_1-h_2)\|}{\|h_1-h_2\|}\right\}\|h_1-h_2\|.$$

The quantity

$$\left\{\|T'(x_0+h_2)-T'(x_0)\| + \frac{\|R_1(x_0+h_2,h_1-h_2)\|}{\|h_1-h_2\|}\right\} \to 0$$

as $\|h_1\| \to 0$ and $\|h_2\| \to 0$. Now we have

$$\|F_\delta(h_1)-F_\delta(h_2)\| \le |\delta|\,\|h_1-h_2\| + \|R_1(x_0,h_1) - R_1(x_0,h_2)\|$$

and the lemma immediately follows.

The following result depends upon the previous lemma:

Theorem 2.2: There exist positive constants c,d such that for $|\delta| < c$, the mapping $h^* = MF_\delta(h)$ carries the ball $\|h\| \le d$ into itself, and is contracting thereon.

Proof: We have

$$\|h^*\| \le \|M\| \left\{ |\delta|\|x_o\| + |\delta|\|h\| + \frac{\|R_1(x_o,h)\|}{\|h\|}\|h\| \right\}.$$

First let us take $d_1 > 0$ small enough that $\frac{\|R_1(x_o,h)\|}{\|h\|} < \frac{1}{2\|M\|}$ for $\|h\| \le d_1$.

Next we can find $c_1 > 0$ so small that

$$|\delta|\|x_o\| + |\delta|\|h\| \le |\delta|(\|x_o\|+d_1) < \frac{d_1}{2\|M\|} \quad \text{for } |\delta| \le c_1.$$

Then $\|h^*\| \le \frac{1}{2} d_1 + \frac{1}{2} d_1 = d_1$, which shows that for $|\delta| < c_1$, $MF_\delta(h)$ maps the ball $\|h\| \le d_1$ into itself.

Again,

$$\|h_1^*-h_2^*\| \le \|M\|\cdot\|F_\delta(h_1) - F_\delta(h_2)\| \le \|M\|\cdot A(\delta,h_1,h_2)\|h_1-h_2\|$$

where we have used the Lipschitz condition satisfied by $F_\delta(h)$. Employing Lemma 2.1, we can take positive constants c_2,d_2 small enough that $\|M\|\cdot A(\delta,h_1,h_2) < \frac{1}{2}$ when $|\delta| < \delta_2$, $\|h_1\| \le d_2$, $\|h_2\| \le d_2$. Then $\|h_1^*-h_2^*\| \le \frac{1}{2}\|h_1-h_2\|$ for $|\delta| < \delta_2$, $\|h_1\| \le d_2$, $\|h_2\| \le d_2$, which shows that for $|\delta| < \delta_2$, $MF_\delta(h)$ is contracting on the ball $\|h\| \le d_2$.

Now if we take $d = d_1 \le d_2$ and let $c = \min(c_1,c_2)$, then $MF_\delta(h)$ maps $\|h\| \le d$ into itself and is also contracting thereon, provided $|\delta| < c$. This proves the theorem.

From the above result, we get the fundamental theorem on extension of solutions:

<u>Theorem 2.3</u>: Suppose that $[\lambda_0 I - T'(x_0)]^{-1}$ exists and is bounded, where x_0, λ_0, $x_0 \in X$, is a solution pair for eq. (2.1). Then there exist positive constants c,d, and a solution $h \in X$ of eq. (2.2), unique in the ball $\|h\| \le d$, provided $|\delta| < c$. Thus the pair x_0+h, $\lambda_0+\delta$ solves eq. (2.1). The constants c,d can be taken so that the operator $(\lambda_0+\delta)I - T'(x_0+h)$ has a bounded inverse for $\|h\| \le d$, $|\delta| < c$. The function $h(\delta)$ solving eq. (2.2) is continuous, and $\lim_{\delta \to 0} h(\delta) = \theta$ where θ is the null element.

<u>Proof</u>: Let c_1, d_1 be the constants of Theorem 2.2. Use of Theorem 2.2 and the Contraction Mapping Principle [ref. 15, p. 27] yields the existence and uniqueness of $h(\delta) \in X$, $|\delta| < \delta_1$, i.e., the solution of eq. (2.3). Since eq. (2.2) and eq. (2.3) are equivalent, the pair δ, h solves eq. (2.2) and the pair x_0+h, $\lambda_0+\delta$ solves eq. (2.1). By a known result [ref. 15, p. 92, Th. 3′] there exist positive constants c_2, d_2 such that $(\lambda_0+\delta)I - T'(x_0+g)$ has a bounded inverse provided $|\delta| < c_2$, $\|g\| \le d_2$. If we take $c = \min(c_1, c_2)$ and $d = \min(d_1, d_2)$, then for $|\delta| < c$ there exists a solution pair $x_0+h(\delta)$, $\lambda_0 + \delta$ satisfying eq. (2.1), unique in the ball $\|h\| \le d$, and such that $(\lambda_0+\delta)I - T'(x_0+h(\delta))$ has a bounded inverse.

Given two values δ, δ^* with $|\delta| < c$, $|\delta^*| < c$, we write

$$\begin{aligned}\|h(\delta)-h(\delta^*)\| &= \|MF_\delta(h(\delta)) - MF_{\delta^*}(h(\delta^*))\| \\ &\le \|MF_\delta(h(\delta)) - MF_{\delta^*}(h(\delta))\| + \|MF_{\delta^*}(h(\delta)) - MF_{\delta^*}(h(\delta^*))\| \\ &\le \|MF_\delta(h(s)) - MF_{\delta^*}(h(\delta))\| + \tfrac{1}{2}\|h(\delta)-h(\delta^*)\|\end{aligned}$$

by the Lipschitz condition derived in the proof of Theorem 2.2 and certainly valid here. Then

$$\|h(\delta)-h(\delta^*)\| \leq 2\|M\|\cdot\|F_\delta(h(\delta)) - F_{\delta^*}(h(\delta))\|$$
$$\leq 2\|M\|\cdot\|x_0+d\|\cdot|\delta-\delta^*|$$

whence $h(\delta)$, $|\delta| < c$, is continuous. The theorem is now proven.

We now study how a solution of eq. (2.1), $x(\lambda)$, valid in a neighborhood of $x_0 = x(\lambda_0)$ by Theorem 2.3, might be extended into the large. We introduce notions developed by T. H. Hildebrandt and L. M. Graves, [ref. 13, sec. 18, p. 151].

We consider the cartesian product $W = X \times R$, where R is the real number system; for $w \in W$, we denote the norm by $\|w\| = \|x\| + |\lambda|$ where $x \in X$, $\lambda \in R$ are the respective components. A neighborhood $N_a(w_0)$ of $w_0 \in W$ consists of points such that $\|x-x_0\| + |\lambda-\lambda_0| < a$ while a neighborhood $N_b(\lambda_0)$ of $\lambda_0 \in R$ comprises points such that $|\lambda-\lambda_0| < b$.

A set $W^o \subset W$ of points $w \in W$ is called a "sheet of points" if it has the following properties:

1) For all $w_0 \in W^{(o)}$, there exist positive constants, a and $b < a$, such that no two points $w_1, w_2 \in N_a(w_0)$ have the same projection $\lambda_* \in R$, i.e., if $w_1=(x_1,\lambda_1)$, $w_2=(x_2,\lambda_2)$, $w_1,w_2 \in N_a(w_0)$, then $\lambda_1 \neq \lambda_2$, and every point $\lambda \in N_b(\lambda_0)$, where $w_0=(x_0,\lambda_0)$, is the projection of a point $w \in W^{(o)}$ contained in $N_a(w_0)$.
2) $W^{(o)}$ is a connected set.

A boundary point w_B of $W^{(o)}$ is a point not belonging to $W^{(o)}$ but such that every neighborhood contains points of $W^{(o)}$, i.e., $w_B \notin W^{(o)}$ but $N_\epsilon(w_B) \cap W^{(o)} \neq 0$, $\epsilon > 0$. Thus $W^{(o)}$ contains only interior points.

A point $w \in W$, $w = (x,\lambda)$ is called an ordinary point with respect to the nonlinear transformation T if $\lambda I-T'(x)$ has a bounded inverse; here

$T'(x)$ is the Fréchet derivative of T at x, [ref. 15, p. 183]. Otherwise w is called an exceptional point of T.

$W^{(o)}$ is called a sheet of solutions of the equation (2.1): $\lambda x = T(x)+f$, $x \in X$, $f \in X$, $\lambda \in R$, if every $w = (x,\lambda)$ in $W^{(o)}$ satisfies $\lambda x = T(x)+f$.

The following theorem is due essentially to Hildebrandt and Graves [ref. 13, p. 152].

Theorem 2.4: If $w_o = (x_o,\lambda_o)$ is an ordinary point with respect to the continuous nonlinear transformation T, i.e. $\lambda_o I-T'(x_o)$ has a bounded inverse, and if $w_o = (x_o,\lambda_o)$ is a solution of eq. (2.1), i.e. $\lambda_o x_o = T(x_o)+f$, then there exists a unique sheet $W^{(o)}$ of solutions with the following properties:

a) $W^{(o)}$ contains w_o.

b) Every point of $W^{(o)}$ is an ordinary point of T.

c) The only boundary points (x_B,λ_B) of the sheet $W^{(o)}$ are exceptional points of T, i.e., $\lambda_B I-T'(x_B)$ does not have a bounded inverse.

Proof: According to Theorem 2.3, there exists at least one sheet of solutions $W^{(1)}$ having properties a) and b). Let $W^{(o)}$ be the "least common superclass" of all such sheets $W^{(1)}$. Evidently $W^{(o)}$ is a connected set of solutions satisfying a) and b). That $W^{(o)}$ is a sheet of solutions of eq. (2.1) follows from Theorem 2.3 and property b).

To show that $W^{(o)}$ satisfies property c), let $w_1 = (x_1,\lambda_1)$ be a boundary point of $W^{(o)}$ and an ordinary point of T. Since T would then be continuous at x_1, $\lambda_1 x_1 = T(x_1)+f$, i.e. w_1 is a solution of equation (2.1). Then however by Theorem 2.3, we could extend $W^{(o)}$ to include w_1 in such a way that the newly extended sheet satisfies a) and b), contradicting the definition of $W^{(o)}$.

Now suppose there is a second sheet $W^{(2)}$ of solutions of equation (2.1) having properties a), b) and c). Then $W^{(2)} \subset W^{(0)}$ and there exists an element $w_1 \in W^{(0)}$ with $w_1 \notin W^{(2)}$. Since $W^{(0)}$ is connected, there exists a continuous function F on R to $W^{(0)}$ such that $F(r_0) = w_0$, $F(r_1) = w_1$, and $r_0 < r_1$. By property a) of $W^{(2)}$, $F(r_0) \in W^{(2)}$. Let r_2=l.u.b.$[r | r_0 \leq r \leq r_1, F(r) \in W^{(2)}]$. Then $F(r_2)$ is a boundary point of $W^{(2)}$. But since $F(r_2) \in W^{(0)}$, it is an ordinary point of T, which contradicts property c) of $W^{(2)}$. This completes the proof.

Every sheet of solutions determines a single valued function $x(\lambda)$ in a neighborhood of each of its points. By Theorem 2.3 these functions are continuous.

The sheet of solutions of Theorem 2.4 is called the "unique maximal sheet" of solutions of eq. (2.1) passing through $w_0 = (x_0, \lambda_0)$. As indicated, the only way for a process of continuation of a branch of solutions $x(\lambda)$ to come to an end is in an approach to a point x_B, λ_B where $\lambda_B I - T'(x_B)$ has no bounded inverse; this is merely an alternative way of saying that any boundary point w_B possessed by a unique maximal sheet of solutions of eq. (2.1) is an exceptional point.

3. Development of Branches of Solutions for Nonlinear Equations near an Exceptional Point. Bifurcation Theory.

Again, as in Section 2, we consider the general bounded continuously Fréchet-differentiable transformation $T(x)$: $X \to X$, with $T(\theta) = \theta$, $\theta \in X$ being the null element. Again we consider solutions of eq. (2.1). Since X is a real space, we stress that we seek real solutions.

Let $x_0 \in X$ be a solution of eq. (2.1) corresponding to $\lambda = \lambda_0$, and consider again the matter of finding nearby solutions; we are immediately led to eqs. (2.2) and (2.3) to be solved for the increment $h \in X$, for given δ. Now however we assume that λ_0 is an exceptional point of T; i.e. $\lambda_0 \in \sigma T'(x_0)$. (See [ref. 21, p. 292] for the spectral notations $\sigma, C\sigma, R\sigma$, and $P\sigma$.)

At the present state of the art, we cannot speak on the behavior of solutions of eq. (2.1) in a neighborhood of x_0 if $\lambda_0 \in C\sigma T'(x_0)$ or if $\lambda_0 \in R\sigma T'(x_0)$. We are equipped only to handle the case $\lambda_0 \in P\sigma T'(x_0)$. Therefore it helps at this point to make the following assumption:

H-1: $T'(x_0)$ is a compact linear operator.

Actually if $T(x)$ is compact and continuous on X (i.e. completely continuous), then by a known theorem [ref. 14, p. 135, Lemma 4.1] the Fréchet derivative $T'(x)$ is also compact, $x \in X$. Thus H-1 is quite a natural assumption.

With $T'(x_0)$ compact, the eigenvalue λ_0 is of finite index ν, the generalized nullspaces $\eta_n(x_0) \subset \eta_{n+1}(x_0)$, $n = 0,1,\dots,\nu-1$ are of finite dimension, and the generalized range $R_\nu(x_0)$ is such that $X = \eta_\nu(x_0) \oplus R_\nu(x_0)$, [ref. 19, p. 183, p. 217]. Thus the null space $\eta_1(x_0)$ and range $R_1(x_0)$ of $\lambda_0 I - T'(x_0)$ each admit the projections E_η and E_R respectively, [ref. 16, problem 1, p. 72].

Since $\lambda_0 \in P\sigma T'(x_0)$, $\lambda_0 I-T'(x_0)$ has no inverse; nevertheless because of the existence of the projection E_η of X on $\eta_1(x_0)$ and the fact that $\lambda_0 I-T'(x_0)$ has a closed range $R_1(x_0)$ (E_R exists), we do have a pseudo-inverse. A pseudo-inverse is a bounded right inverse defined on the range: $[\lambda_0 I-T'(x_0)]Mx = x$, $x \in R_1(x_0)$.

We state and prove the following lemma about the pseudo-inverse, which is applicable here [ref. 16, p. 72]:

Lemma 3.1: Let A be a closed linear operator and suppose R(A) is closed. If $\eta(A)$ admits a projection E then A has a pseudo-inverse. Conversely if $\mathcal{D}(A) = X$ and A has a pseudo-inverse, then $\eta(A)$ admits a projection. Here of course R(A), $\eta(A)$ and $\mathcal{D}(A)$ stand for range of A, nullspace of A and domain of A respectively.

Proof. The operator $\hat{A}$ induced by A on $X/\eta(A)$ is 1:1 onto R(A), and thus has a bounded inverse. Therefore $\|\hat{A}[x]\| \geq \gamma\|[x]\| > \frac{1}{2}\gamma \inf_{z\in[x]}\|z\|$ for any $[x]\in X/\eta(A)$, where γ is the minimum modulus of A, [ref. 10, p. 96]. Hence given $y \in R(A)$ there exists an element $x \in [x]$ with $y = Ax$ such that $\|x\| \leq c\|Ax\| = c\|y\|$, where $c = \frac{2}{\gamma}$.

Now define M on R(A) as follows: put $My = (I-E)x$ where $y \in R(A)$, $y = Ax$ and E projects on $\eta(A)$. M is well defined; indeed if $y = Ax_1 = Ax_2$, then $x_1-x_2 \in \eta(A)$, whence $(I-E)x_1 = (I-E)x_2$. Also $AM = I$ since $AMy = A(I-E)x = Ax = y$, $y \in R(A)$, and M is bounded: $\|My\| = \|(I-E)x\| \leq K_1\|x\| \leq cK_1\|y\|$, using a proper choice of x.

On the other hand, if $\mathcal{D}(A) = X$, let M be the given pseudo-inverse. A is bounded by the Closed Graph theorem. Therefore $E = I-MA$ is bounded. Since $AEx = 0$, $R(E) \subset \eta(A)$. If $x \in \eta(A)$ then $Ex = (I-MA)x = x$. Hence E is the projection on $\eta(A)$, and the lemma is proven.

Henceforth, let $M(x_o)$ be the pseudo-inverse of $\lambda_o I - T'(x_o)$ given by the lemma. We have

$$[\lambda_o I - T'(x_o)]M(x_o) = I \quad \text{on} \quad R_1(x_o)$$

$$M(x_o)[\lambda_o I - T'(x_o)] = E_R.$$

We extend the pseudo-inverse $M(x_o)$ to the entire space X by writing $\overline{M}(x_o) = M(x_o)E_R$. Then

$$[\lambda_o I - T'(x_o)]\overline{M}(x_o) = \overline{M}(x_o)[\lambda_o I - T'(x_o)] = E_R. \tag{3.1}$$

With the aid of the extended pseudo-inverse, let us study the following equation to be solved for h:

$$h = \overline{M}(x_o)F_\delta(h) + u, \qquad u \in \eta_1(x_o) \tag{3.2}$$

where as before (see eq. (2.4))

$$F_\delta(h) = -\delta x_o - \delta h + R_1(x_o,h). \tag{3.3}$$

If $h \in X$ satisfies eq. (2.3) for given x_o,λ_o,δ, then $F_\delta(h) \in R_1(x_o)$. Using eq. (3.1) we see that $u = h - \overline{M}(x_o)F_\delta(h) \in \eta_1(x_o)$, so that the same h satisfies eq. (3.2) with this u. Therefore we are motivated to prove an existence theorem for eq. (3.2):

Theorem 3.2: There exist positive constants c,d,e such that for $|\delta| < c$ and $\|u\| < e$, $u \in \eta_1(x_o)$, eq. (3.2) has a solution $h(\delta,u)$ unique in the ball $\|h\| \le d$. The solution is continuous in δ.

Proof: We study the mapping $h^* = \overline{M}(x_o)F_\delta(h) + u$ of X into itself, $u \in \eta_1(x_o)$. We have

$$\|h^*\| \le \|\overline{M}\| \left\{ |\delta|\,\|x_o\| + |\delta|\cdot\|h\| + \frac{\|R_1(x_o,h)\|}{\|h\|}\cdot\|h\| \right\} + \|u\|$$

according to our definition (3.3) of $F_\delta(h)$. First we can take d_1 so small that $\frac{\|R_1(x_0,h)\|}{\|h\|} \le \frac{1}{3\|\bar{M}\|}$ for $\|h\| < d_1$. With d_1 thus fixed, we can find δ_1 such that

$$|\delta|\cdot\|x_0\| + |\delta|\cdot\|h\| \le |\delta|(\|x_0\|+d_1) \le \frac{d_1}{3\|\bar{M}\|} \quad \text{for} \quad |\delta| < \delta_1.$$

Next we take $\|u\| \le \frac{d_1}{3}$; then $\|h^*\| = \|\bar{M}\, F_\delta(h)+u\| \le d_1$ if $|\delta| < \delta_1$. Thus if $|\delta| < \delta_1$, and $\|u\| < \frac{d_1}{3}$ the map carries the ball $\|h\| < d_1$ into itself.

In view of Lemma 2.1 we can find d_2,δ_2 small enough in order to have $\|\bar{M}\|A(\delta,h_1,h_2) < \frac{1}{2}$ for $|\delta| < \delta_2$, $\|h_1\| \le d_2$, $\|h_2\| \le d_2$. Thus $\|h_1{}^*-h_2{}^*\| = \|\bar{M}F_\delta(h_1)-\bar{M}F_\delta(h_2)\| < \frac{1}{2}\|h_1-h_2\|$ provided $|\delta|<\delta_2$, $\|h_1\|\le d_2$, $\|h_2\| \le d_2$. Take $c = \min(c_1,c_2)$, $d = d_1 \le d_2$, and $e = \frac{d}{3}$. Then the map carries the ball $\|h\| \le d$ into itself and is contracting thereon, provided $\|u\| \le e$. Therefore if $|\delta| < c$ and $\|u\| \le e$, the iterations $h_{n+1} = \bar{M}F_\delta(h_n) + u$ converge to a solution of eq. (3.2) unique in the ball $\|h\| < d$. The continuity is obtained in a way similar to that used in Theorem 2.3. This ends the proof.

The existence and local uniqueness of a solution of eq. (3.2) given in Theorem 3.2 sets the stage for the following result:

<u>Theorem 3.3</u>: Let $T'(x_0)$ satisfy H-1, (which can be arranged by assuming that $T(x)$ is everywhere compact and continuous). Then the condition that $h = V_\delta(u)$, the solution of eq. (3.2), be at the same time a solution of eq. (2.3) is that

$$(I-E_R)F_\delta(V_\delta(u)) = \theta \tag{3.4}$$

where of course E_R is the projection of X onto the range $R_1(x_o)$. Conversely if h,δ satisfy eq. (2.3), then $F_\delta(h) \in R_1(x_o)$, eq. (3.4) is satisfied, and h,δ also satisfy eq. (3.2).

Proof: Since $T'(x_o)$ is compact, $\lambda_o I-T'(x_o)$ has closed range, and as we have seen, the null space $\eta_1(x_o)$ admits the projection E_η. Thus by Lemma 3.1 the pseudo-inverse M and the extended pseudo-inverse $\overline{M} = ME_R$ exist where E_R projects on the range $R_1(x_o)$. We have $\mathcal{D}(\overline{M}) = X$. Let δ and $u \in \eta_1(x_o)$ be such that eq. (3.4) is satisfied, where $V_\delta(u)$ is the solution of eq. (3.2). Then $F_\delta(V_\delta(u)) \in R_1(x_o) = \mathcal{D}(M)$. Premultiplication of eq. (3.2) with $\lambda_o I-T'(x_o)$ and use of eq. (3.1) give

$$(\lambda_o I-T'(x_o))h = (\lambda_o I-T'(x_o))\overline{M}F_\delta(h) + \theta = F_\delta(h),$$

which is just eq. (2.3). On the other hand if h,δ satisfies eq. (2.3), then $F_\delta(h) \in R_1(x_o)$. Let $u_o = h - \overline{M}F_\delta(h)$; then

$$[\lambda_o I-T'(x_o)]u_o = [\lambda_o I-T'(x_o)]h-[\lambda_o I-T'(x_o)]\overline{M}F_\delta(h)$$

$$= F_\delta(h) - E_R F_\delta(h) = \theta$$

by eq. (3.1). Thus $h = \overline{M}F_\delta(h) + u_o$ with $u_o \in \eta_1(x_o)$, which is eq. (3.2). Then $h = V_\delta(u_o)$, and since $F_\delta(h) \in R_1(x_o)$, we have $(I-E_R)F_\delta(V_\delta(u_o)) = \theta$, which is eq. (3.4). This ends the proof.

In the proof of Theorem 3.2, the solution of eq. (3.2) was produced by the method of contraction mappings under the assumption that $|\delta| < c$ and $\|u\| \le e$, where c and e are positive constants. Hence $h = V_\delta(u)$ is the unique limit of the iterations $h_{n+1} = \overline{M}F_\delta(h_n) + u$, which converge in norm. We compose these iterates to get the nonlinear expansion:

$$V_\delta(u) = u + \overline{M}F_\delta[u+\overline{M}F_\delta[u+\overline{M}F_\delta[u+\cdots etc.]]]. \tag{3.5}$$

By Theorems 3.2 and 3.3, in order to study small solutions h,δ of eq. (2.3), it is well to study small solutions of eq. (3.4) and eq. (3.2). Eq. (3.4) is to be solved for $u \in \eta_1(x_o)$, where $\eta_1(x_o)$ is of course a finite dimensional subspace of X. By a known result [ref. 21, p. 285, Th. 1], the annihilator of the range $R_1(x_o)$ is the null space $\eta_1{}^*(x_o)$ of the adjoint $\lambda_o I-T'(x_o)^*$. Thus by choosing bases in $\eta_1(x_o)$ and $\eta_1{}^*(x_o)$, eq. (3.4) may be regarded as a finite system of nonlinear scalar equations in an equal finite number of scalar unknowns, parameterized by the real scalar δ. Indeed if $u_1,\cdots,u_n$ and $u_1{}^*,\cdots,u_n{}^*$ are bases respectively for $\eta_1(x_o)$ and $\eta_1{}^*(x_o)$, then eq. (3.4) has the representation

$$u_i{}^*F_\delta(V_\delta(\xi_o u_1+\cdots+\xi_n u_n)) = 0 \tag{3.6}$$
$$i = 1,2,\cdots,n.$$

This system of nonlinear scalar equations is to be solved for the scalar unknowns $\xi_j(\delta)$, $j = 1,2,\cdots,n$. (Note: $u_i{}^* \in X^*$, $i = 1,\cdots,n$.)

Before proceeding further with solution of eq. (3.4), it is necessary to make it more amenable to calculation. First we must expand $F_\delta(h)$ to more terms:

$$F_\delta(h) = -\delta x_o - \delta h + \frac{1}{2!}d^2T(x_o;h,h) + \frac{1}{3!}d^3T(x_o;h,h,h) + R_3(x_o,h) \tag{3.7}$$

where $d^2T(x_o;h_1,h_2)$ and $d^3T(x_o;h_1,h_2,h_3)$ are respectively the second and third Fréchet differentials of the nonlinear operator $T(x)$, linear in each argument h_i separately, and symmetric in the arguments h_i, [ref. 15, p. 188]. The remainder $R_3(x_o,h)$ is of course such that

$$\frac{\|R_3(x_o,h)\|}{\|h\|^3} \to 0 \text{ as } \|h\| \to 0.$$

Following R. G. Bartle [ref. 1, p. 370, 373], we now substitute the nonlinear expansion, eq. (3.5), into the expression, eq. (3.7), for $F_\delta(h)$ to compute $F_\delta(V_\delta(u))$. As a first step we write

$$\begin{aligned}
F_\delta(V_\delta(u)) = &- \delta x_o - \delta[u+\bar{M}F_\delta(h)] + \frac{1}{2!} d^2T(x_o;u+\bar{M}F_\delta(h), u + \bar{M}F_\delta(h)) \\
&+ \frac{1}{3!} d^3T(x_o;u+\bar{M}F_\delta(h), u + \bar{M}F_\delta(h), u + \bar{M}F_\delta(h)) \\
&+ R_3(x_o,u+\bar{M}F_\delta(h)) \\
= &- \delta x_o - \delta u - \delta\bar{M}F_\delta(h) + \frac{1}{2!} d^2T(x_o;u,u) \\
&+ 2\cdot\frac{1}{2!} d^2T(x_o;u,\bar{M}F_\delta(h)) + \frac{1}{2!} d^2T(x_o;\bar{M}F_\delta(h), \bar{M}F_\delta(h)) \\
&+ \frac{1}{3!} d^3T(x_o;u,u,u) + 3\cdot \frac{1}{3!} d^3T(x_o;u,u,\bar{M}F_\delta(h)) \\
&+ 3\cdot \frac{1}{3!} d^3T(x_o;u,\bar{M}F_\delta(h), \bar{M}F_\delta(h)) \\
&+ \frac{1}{3!} d^3T(x_o;\bar{M}F_\delta(h), \bar{M}F_\delta(h), \bar{M}F_\delta(h)) + R_3(x_o,u+\bar{M}F_\delta(h)).
\end{aligned}$$

Continuing to substitute,

$$\begin{aligned}
F_\delta(V_\delta(u)) = &- \delta x_o - \delta u + \frac{1}{2!} d^2T(x_o;u,u) + \frac{1}{3!} d^3T(x_o;u,u,u) \\
&- \bar{M}\left\{- \delta^2 x_o - \delta^2 h + \frac{\delta}{2!} d^2T(x_o;h,h) + \frac{\delta}{3!} d^3T(x_o;h,h,h) + \delta R_3(x_o,h)\right\} \\
&+ \frac{2}{2!} d^2T(x_o;u,\bar{M}[-\delta x_o-\delta h+\frac{1}{2!} d^2T(x_o;h,h) + \frac{1}{3!} d^3T(x_o;h,h,h) + R_3(x_o,h)]) \\
&+ \frac{1}{2!} d^2T(x_o;\bar{M}[-\delta x_o-\delta h+ \frac{1}{2!} d^2T(x_o;h,h) + \frac{1}{3!} d^3T(x_o;h,h,h) + R_3(x_o,h)], \\
&\qquad \bar{M}[-\delta x_o-\delta h + \frac{1}{2!} d^2T(x_o;h,h) + \frac{1}{3!} d^3T(x_o;h,h,h) + R_3(x_o,h)]) \\
&+ \frac{3}{3!} d^3T(x_o;u,u,\bar{M}[-\delta x_o-\delta h + \frac{1}{2!} d^2T(x_o;h,h)+ \frac{1}{3!} d^3T(x_o;h,h,h)+R_3(x_o,h)])
\end{aligned}$$

$$+ \frac{3}{3!} d^3T(x_o;u,\overline{M}[-\delta x_o-\delta h + \frac{1}{2!} d^2T(x_o;h,h) + \frac{1}{3!} d^3T(x_o;h,h,h) + R_3(x_o,h)],$$

$$\overline{M}[-\delta x_o-\delta h + \frac{1}{2!} d^2T(x_o;h,h) + \frac{1}{3!} d^3T(x_o;h,h,h) + R_3(x_o,h)])$$

$$+ \frac{1}{3!} d^3T(x_o;\overline{M}[-\delta x_o-\delta h + \frac{1}{2!} d^2T(x_o;h,h) + \frac{1}{3!} d^3T(x_o;h,h,h) + R_3(x_o,h)],$$

$$\overline{M}[-\delta x_o-\delta h + \frac{1}{2!} d^2T(x_o;h,h) + \frac{1}{3!} d^3T(x_o;h,h,h) + R_3(x_o,h)],$$

$$\overline{M}[-\delta x_o-\delta h + \frac{1}{2!} d^2T(x_o;h,h) + \frac{1}{3!} d^3T(x_o;h,h,h) + R_3(x_o,h)])$$

$$+ R_3(x_o,u+\overline{M}[-\delta x_o-\delta h + \frac{1}{2!} d^2T(x_o;h,h) + \frac{1}{3!} d^3T(x_o;h,h,h) + R_3(x_o,h)]).$$

Since eq. (3.7) contains no terms in which h appears alone raised to the first power, this process results in terms containing h and δ to successively higher degrees.

So as to estimate the higher order terms, consider eq. (3.2):

$$h = \overline{M}(x_o)F_\delta(h) + u$$

$$= \overline{M}(x_o)\{-\delta x_o-\delta h + \frac{1}{2!} d^2T(x_o;h,h) + \frac{1}{3!} d^3T(x_o;h,h,h)$$

$$+ R_3(x_o,h)\} + u = \overline{M}(x_o)F_\delta{}^o(h) - \delta\overline{M}(x_o)x_o + u,$$

where $F_\delta{}^o(h) = -\delta h + \frac{1}{2!} d^2T(x_o;h,h) + \frac{1}{3!} d^3T(x_o;h,h,h) + R_3(x_o,h)$.

Note that $F_\delta{}^o(\theta) = \theta$ and

$$\|\overline{M}(x_o)\|\cdot\|F_\delta{}^o(h_1) - F_\delta{}^o(h_2)\| \le \frac{1}{2}\|h_1-h_2\|, \quad \|h_1\| \le d_2, \quad \|h_2\| \le d_2, \quad |\delta| \le \delta_2,$$

where d_2, δ_2 are numbers arising in the proof of Theorem 3.2.

Thus $\|\overline{M}(x_o)\|\cdot\|F_\delta{}^o(h)\| \le \frac{1}{2}\|h\|$, and

$$\|h\| \le \|\overline{M}(x_o)\|\cdot\|F_\delta{}^o(h)\| + |\delta|\cdot\|\overline{M}(x_o)\|\cdot\|x_o\| + \|u\|$$

$$\le \frac{1}{2}\|h\| + c|\delta| + \|u\|, \text{ where } c = \|\overline{M}(x_o)\|\cdot\|x_o\|,$$

or

$$\|h\| \le 2(c|\delta| + \|u\|), \qquad \|h\| \le d_2, \quad |\delta| \le \delta_2.$$

Thus to estimate the terms in u and δ which eventually arise, we note that

$$\|d^3T(x_0;h,h,h)\| \le \text{const } \|h\|^3 \le \text{const } (c|\delta| + \|u\|)^3,$$

and so

$$d^3T(x_0;h,h,h) = O([c|\delta| + \|u\|]^3),$$

and similarly for other expressions. We propose to keep explicit for the time being only those terms of order up to, but not including, $O(|\delta|^i\|u\|^j)$, $i + j = 3$.

Now,

$$\begin{aligned} R_3(x_0,h) &= d^4T(x_0+t_0h;h,h,h,h), \quad 0 \le t_0 \le 1, \\ &= O([c|\delta| + \|u\|]^4), \end{aligned}$$

so that we do not keep the remainders explicit.

In the important case of bifurcation at the origin, $x_0 = \theta$, so that $c = 0$.

Lumping the terms that we do not keep explicit, we have

$$\begin{aligned} F_\delta(V_\delta(u)) = &- \delta x_0 - \delta u + \frac{1}{2!} d^2T(x_0;u,u) + \frac{1}{3!} d^3T(x_0;u,u,u) \\ &- \overline{M}(x_0)\{-\delta^2 x_0 - \delta^2 h + \frac{\delta}{2!} d^2T(x_0;h,h)\} \\ &+ \frac{2}{2!} d^2T(x_0;u,\overline{M}\{-\delta x_0 - \delta h + \frac{1}{2!} d^2T(x_0;h,h)\}) \\ &+ \frac{1}{2!} d^2T(x_0;\overline{M}(x_0)\{-\delta x_0 - \delta h + \frac{1}{2!} dT(x_0;h,h)\},\overline{M}(x_0)\{-\delta x_0 - \delta h + \frac{1}{2!} d^2T(x_0;h,h)\}) \\ &+ \frac{3}{3!} d^3T(x_0;u,u,\overline{M}(x_0)\{-\delta x_0\}) + \frac{3}{3!} d^3T(x_0;u,\overline{M}(x_0)\{-\delta x_0\}, \overline{M}(x_0)\{-\delta x_o\}) \end{aligned}$$

$$+ \frac{1}{3!} d^3T(x_0;\bar{M}\{-\delta x_0\},\ \bar{M}\{-\delta x_0\},\ \bar{M}\{-\delta x_0\})$$

$$+ \sum_{i+j=3} \omega_{ij}(\delta,u) \quad \text{where} \quad \omega_{ij}(\delta,u) = o(|\delta|^i\|u\|^j)$$

$$= -\delta x_0 - \delta u + \frac{1}{2!} d^2T(x_0;u,u) + \frac{1}{3!} d^3T(x_0;u,u,u)$$

$$+ \delta^2\bar{M}x_0 + \delta^2\bar{M}h - \frac{\delta\bar{M}}{2!} d^2T(x_0;h,h) - \delta d^2T(x_0;u,\bar{M}x_0)$$

$$- \delta d^2T(x_0;u,\bar{M}h) + \frac{1}{2!} d^2T(x_0;u,d^2T(x_0;h,h))$$

$$+ \frac{\delta^2}{2!} d^2T(x_0;\bar{M}x_0,\bar{M}x_0) + \delta^2 d^2T(x_0;\bar{M}h,\bar{M}x_0)$$

$$- \frac{\delta}{2!} d^2T(x_0;\bar{M}x_0,\bar{M}d^2T(x_0;h,h))$$

$$- \frac{\delta}{2!} d^3T(x_0;u,u,\bar{M}x_0) + \frac{\delta^2}{2!} d^3T(x_0;u,\bar{M}x_0,\bar{M}x_0)$$

$$- \frac{\delta^3}{3!} d^3T(x_0;\bar{M}x_0,\bar{M}x_0,\bar{M}x_0) + \sum_{i+j=3} \omega_{ij}(\delta,u)$$

$$\text{where } \omega_{ij}(\delta,u) = o(|\delta|^i\|u\|^j).$$

Again, substituting $h = u + \bar{M}F_\delta(h)$, we finally get

$$F_\delta(V_\delta(u)) = -\delta x_0 - \delta u + \frac{1}{2} d^2T(x_0;u,u) + \frac{1}{6} d^3T(x_0;u,u,u)$$

$$+ \delta^2\bar{M}x_0 + \delta^2\bar{M}u - \delta^3\bar{M}^2x_0 - \frac{\delta\bar{M}}{2} d^2T(x_0;u,u)$$

$$+ \delta^2\bar{M}d^2T(x_0;u,\bar{M}x_0) - \frac{\delta^3\bar{M}}{2} d^2T(x_0;\bar{M}x_0,\bar{M}x_0)$$

$$- \delta d^2T(x_0;u,\bar{M}x_0) - \delta d^2T(x_0;u,\bar{M}u) + \delta^2 d^2T(x_0;u,\bar{M}^2x_0)$$

$$+ \frac{1}{2} d^2T(x_0;u,d^2T(x_0;u,u)) - \delta d^2T(x_0;u,d^2T(x_0;u,\bar{M}x_0))$$

$$+ \frac{1}{2} \delta^2 d^2T(x_0;u,d^2T(x_0;\bar{M}x_0,\bar{M}x_0))$$

$$+ \frac{\delta^2}{2} d^2T(x_o;\bar{M}x_o,\bar{M}x_o) + \delta^2 d^2T(x_o;\bar{M}u,\bar{M}x_o) - \delta^3 d^2T(x_o;\bar{M}^2x_o,\bar{M}x_o)$$

$$- \frac{\delta}{2} d^2T(x_o;\bar{M}x_o,\bar{M}d^2T(x_o;u,u)) + \delta^2 d^2T(x_o;\bar{M}x_o,\bar{M}d^2T(x_o;u,\bar{M}x_o))$$

$$- \frac{\delta^3}{2} d^2T(x_o;\bar{M}x_o,\bar{M}d^2T(x_o;\bar{M}x_o,\bar{M}x_o))$$

$$- \frac{\delta}{2} d^3T(x_o;u,u,\bar{M}x_o) + \frac{\delta^2}{2} d^3T(x_o;u,\bar{M}x_o,\bar{M}x_o)$$

$$- \frac{\delta^3}{6} d^3T(x_o;\bar{M}x_o,\bar{M}x_o,\bar{M}x_o) + \sum_{i+j=3} \omega_{ij}(\delta,u)$$

$$= - \delta x_o - \delta\{u + d^2T(x_o;u,\bar{M}x_o)\}$$

$$+ \frac{1}{2} d^2T(x_o;u,u) + \frac{1}{6} d^3T(x_o;u,u,u) + \sum_{i+j=1}^{3} \omega_{ij}(\delta,u)$$

where of course $\omega_{ij}(\delta,u) = o(|\delta|^i\|u\|^j)$.

Again if $u_1,\cdots,u_n$ and $u_1^*,\cdots,u_n^*$ are bases respectively for the null spaces $\eta_1(x_o)$and $\eta_1^*(x_o)$, then the bifurcation equation, eq. (3.4) or eq. (3.6), becomes,

$$- \delta u_k^* x_o - \delta \left\{ u_k^* u_k + u_k^* \sum_{\ell=1}^{n} \xi_\ell d^2T(x_o;u_\ell,\bar{M}x_o) \right\}$$

$$+ \frac{1}{2} u_k^* \sum_{\ell_1=1}^{n} \sum_{\ell_2=1}^{n} \xi_{\ell_1}\xi_{\ell_2} d^2T(x_o;u_{\ell_1},u_{\ell_2}) \tag{3.8}$$

$$+ \frac{1}{6} u_k^* \sum_{\ell_1=1}^{n} \sum_{\ell_2=1}^{n} \sum_{\ell_3=1}^{n} \xi_{\ell_1}\xi_{\ell_2}\xi_{\ell_3} d^2T(x_o;u_{\ell_1},u_{\ell_2},u_{\ell_3})$$

$$+ u_k^* \sum_{i+j=1}^{3} \omega_{ij}\left(\delta, \sum_{\ell=1}^{n} u_\ell\right) = 0 \qquad k = 1,2,\cdots,n$$

$$\|u_k\| = \|u_k^*\| = 1.$$

This representation of the bifurcation equation has validity, of course, for those values of δ, ξ_ℓ such that our iterative solution V_δ of eq. (3.2) has validity, i.e., for $|\delta| \le c$, $|\xi_\ell| \le e$, where c,e are constants occurring in Theorem 3.2.

4. Solution of the Bifurcation Equation in the Case n = 1; Bifurcation at the Origin.

Eq. (3.8) represents n equations to be solved for n scalar unknowns $\xi_1,\dots,\xi_n$. If solution can be accomplished, then by Theorem 3.3, letting $u = \sum_{k=1}^{n} \xi_k u_k$ in eq. (3.2), we produce solutions of eq. (2.1) in a neighborhood of the exceptional point (x_o,λ_o). Here, of course, the elements $u_1,\dots,u_n$ are assumed to be a basis for the null space $\eta_1(x_o)$ of the operator $\lambda_o I - T'(x_o)$.

The solution of eqs. (3.8) presents much difficulty in the cases $n > 1$, although attempts have been and are being made to study these cases, [refs. 3, 12 and 48]. For the present, we here confine the discourse to the case $n = 1$.

If we assume the null spaces $\eta_1(x_o)$ and $\eta_1^*(x_o)$ to be one dimensional, and to be spanned by the elements u_1 and u_1^* respectively, i.e., $T'(x_o)u_1 = \lambda_o u_1$, $T'(x_o)^* u_1^* = \lambda_o u_1^*$, then eq. (3.8) simplifies to the following single equation in the scalar unknown ξ_1 (note: $u_1 \in X$, $u_1^* \in X^*$):

$$
\begin{aligned}
-\delta u_1^* x_o - \delta\xi_1 u_1^*[u_1 + d^2T(x_o;u_1,\bar{M}x_o)] \\
+ \frac{\xi_1^2}{2} u_1^* d^2T(x_o;u_1,u_1) + \frac{\xi_1^3}{6} u_1^* d^3T(x_o;u_1,u_1,u_1) \qquad (4.1)\\
+ \sum_{i+j=1}^{3} u_1^* \omega_{ij}(\delta,\xi_1) = 0, \text{ where } \omega_{ij}(\delta,\xi_1) = o(|\delta|^i|\xi_1|^j), \\
\text{and } \|u_1\| = \|u_1^*\| = 1.
\end{aligned}
$$

For convenience we write eq. (4.1) as follows:

$$\delta[a_1+\Phi_1(\delta,\xi_1)] + \xi_1\delta[a_2+\Phi_2(\delta,\xi_1)] + \xi_1^2[a_3+\Phi_3(\delta,\xi_1)] + \xi_1^3[a_4+\Phi_4(\delta,\xi_1)] = 0 \tag{4.2}$$

where $a_1 = - u_1{}^*x_0$ $\qquad a_2 = - u_1{}^*[u_1+d^2T(x_0;u_1,\bar{M}x_0)]$

$a_3 = \frac{1}{2} u_1{}^*d^2T(x_0;u_1,u_1)$ $\qquad a_4 = \frac{1}{6} u_1{}^*d^3T(x_0;u_1,u_1,u_1)$

and we have defined

$$\Phi_1(\delta,\xi_1) = \frac{u_1{}^*(\omega_{10}+\omega_{20}+\omega_{30})}{\delta}, \qquad \Phi_2(\delta,\xi_1) = \frac{u_1{}^*(\omega_{11}+\omega_{12}+\omega_{21})}{\delta\xi_1}$$

$$\Phi_3(\delta,\xi_1) = \frac{u_1{}^*\omega_{02}}{\xi_1^2}, \qquad \Phi_4(\delta,\xi_1) = \frac{u_1{}^*\omega_{03}}{\xi_1^3},$$

and where $\Phi_i(\delta,\xi_1) \to 0$ as $\delta,\xi_1 \to 0$, $i = 1, 2, 3, 4$.

Eq. (4.2) is of the form

$$\sum_{i=1}^{m} \delta^{\alpha_i}\xi_1^{\beta_i}\,[a_i+\Phi_i(\delta,\xi_1)] = 0 \tag{4.3}$$

with $m = 4$; $\alpha_1 = 1$, $\beta_1 = 0$; $\alpha_2 = 1$, $\beta_2 = 1$; $\alpha_3 = 0$, $\beta_3 = 2$; $\alpha_4 = 0$, $\beta_4 = 3$. Equations such as eq. (4.3) were treated by the method of the Newton Polygon by J. Dieudonné [ref. 8] and R. G. Bartle [ref. 1, p. 376]. In each of these studies it was necessary to assume that, among those terms in eq. (4.3) with $a_i \neq 0$, $\min_{1\le i\le n} \alpha_i = \min_{1\le i\le n} \beta_i = 0$. Thus with the exponents listed for eq. (4.2) we should want $a_1 \neq 0$, and $a_3 \neq 0$ or $a_4 \neq 0$ in order to employ the Newton Polygon method as developed by these two authors.

We now take up the study of bifurcation at the origin. In eq. (2.1) we take $f \equiv 0$ so that we consider now the eigenfunctions of the nonlinear operator $T(x)$. In other words, from now on we interest ourselves in the eigenvalue problem $\lambda x = T(x)$, where $T(\theta) = \theta$, the null element. There exists the trivial solution $x = \theta$, and we have the problem of determining those real values of λ such that nontrivial solutions exist. The pair (θ, λ_o) is a solution pair of eq. (2.1); if it happens to be an exceptional point as defined in connection with Theorem 2.4, then we have the problem of bifurcation at the origin.

It is convenient at this time also to assume that the nonlinear operator $T(x)$ is odd: $T(-x) = -T(x)$. Thus

H-2: $T(x)$ is an odd, thrice differentiable operator.

With odd operators we have the following result:

Theorem 4.1: Let the nonlinear operator $T(x)$ satisfy H-2. Then $T'(x)$ is an even operator: $T'(-x) = T'(x)$, and $T''(x)$ is an odd operator: $T''(-x) = -T''(x)$.

Proof: By a known result [ref. 15, p. 185] the weak derivative exists, and we have $T'(-x)h = \lim_{t\to o} \frac{1}{t}[T(-x+th) - T(-x)] = \lim_{t\to o} \frac{1}{(-t)}[T(x-th) - T(x)]$ $= T'(x)h$. This shows the evenness of $T'(x)$. Again $T''(-x)h_1h_2 = d^2T(-x;h_1,h_2)$ $= \lim_{t\to o} \frac{1}{t}[T'(-x+th_1)-T'(-x)]h_2 = \lim_{t\to o} \frac{1}{t}[T'(x-th_1)-T'(x)]h_2 =$ $-\lim_{t\to o} \frac{1}{(-t)}[T'(x-th_1)-T'(x)]h_2 = -T''(x)h_1h_2$, or $T''(-x) = -T''(x)$, which shows the oddness of $T''(x)$. Here, of course, $x,h,h_1,h_2 \in X$. This ends the proof.

Just as the oddness of $T(x)$ implies that $T(\theta) = \theta$, so does the oddness of $T''(x)$ imply that $T''(\theta)h_1h_2 = d^2T(\theta;h_1,h_2) = \theta, h_1,h_2 \in X$. By Theorem 4.1, eq. (4.1) appears as follows:

$$- \delta \xi_1 u_1{*}u_1 + \frac{\xi_1^{\ 3}}{6} u_1{*}d^3T(\theta;u_1,u_1,u_1) + \sum_{i+j=1}^{3} u_1{*}w_{ij}(\delta,\xi_1) = 0.$$

With $x_o = \theta$, we also discern that $R_3(\theta,h) = O(\|u\|^4)$; this, and other terms in the expansion of $F_\delta(V_\delta(u))$ which vanish, imply that $\omega_{10} = \omega_{20} = \omega_{30} = \omega_{02} = \theta$. Hence the bifurcation equation in the case $x_o = \theta$ can be written as follows:

$$\xi_1\delta[a_2+\Phi_2(\delta,\xi_1)] + \xi_1^{\ 3}[a_4+\Phi_4(\delta,\xi_1)] = 0. \qquad (4.4)$$

The coefficients a_2,a_4 and the functions $\Phi_2(\delta,\xi_1)$, $\Phi_4(\delta,\xi_1)$ are as defined in connection with eq. (4.2). It is seen that ξ_1 can be cancelled in eq. (4.4).

At this point we explain the manner in which J. Dieudonné treated equations such as (4.2), (4.3) and (4.4). Let $f(\delta,\xi_1)$ be the left-hand side of either eqs. (4.2), (4.3) or (4.4). If the function $\psi(\delta)$ solves the equation $f(\delta,\xi_1) = 0$, i.e., $f(\delta,\psi(\delta)) \equiv 0$, in the neighborhood of $(0,0)$, $(f(0,0) = 0)$, then $\psi(\delta) \sim t\delta^\mu$, where $-\frac{1}{\mu}$ is the slope of one of the sides of the Newton Polygon, and t is a real root of the equation $\sum_k a_k t^k = 0$. Here k runs over the indices of the points on the side of the polygon of which $-\frac{1}{\mu}$ is the slope, [ref. 8, p. 50]. Conversely, to a given side and slope of the Newton Polygon, there _may_ correspond a solution in the

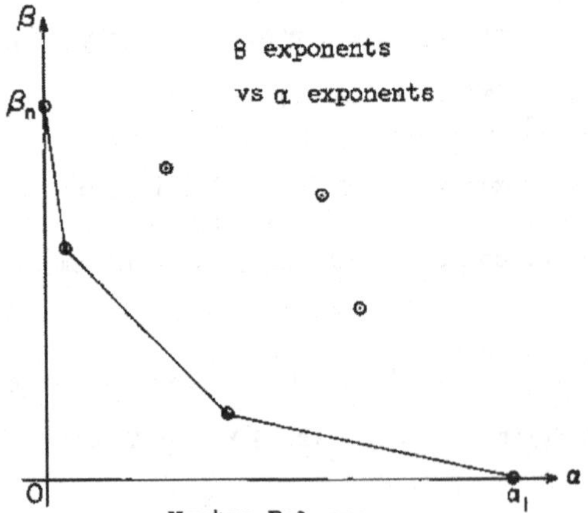

Newton Polygon

FIG. 4.1.

small, $\phi(\delta) \sim t\delta^{\mu}$, of $f(\delta,\xi_1) = 0$.

For simplicity let us take the case where there is only one side, of slope $-\frac{1}{\mu_1}$, of the Newton Polygon. Put $\xi_1 = \eta\delta^{\mu_1}$ in $f(\delta,\xi_1) = 0$. After division by $\delta^{\mu_1\beta_1}$, we have, (cf. eq. (4.3)):

$$\sum_{i=1}^{m} \delta^{\alpha_i+\mu_1\beta_i-\mu_1\beta_1}\eta^{\beta_i}\left[a_i+\Phi_i(\delta,\eta\delta^{\mu_1})\right] = 0.$$

Noting now that $\alpha_k + \mu_1\beta_k - \mu_1\beta_1 = 0$ for all points on the single side of the Newton Polygon, we have

$$\sum_k \eta^{\beta_k}\left[a_k+\Phi_k(\delta,\eta\delta^{\mu_1})\right] + \sum_i \delta^{\alpha_i+\mu_1\beta_i-\mu_1\beta_1}\eta^{\beta_i}\left[a_i+\Phi_i(\delta,\eta\delta^{\mu_1})\right] = 0, \quad (4.5)$$

where the second sum is over all the remaining points.

Now let t_o be a real root of the equation $a_1t^{\beta_1} + \sum_k a_kt^{\beta_k} = 0$, of multiplicity q. Then eq. (4.5) can be written as follows:

$$(\eta-t_o)^q = \delta^{\lambda}F(\delta,\eta), \qquad \lambda > 0,$$

where $F(\delta,\eta)$ is continuous and tends to $b \neq 0$ as $(\delta,\eta) \to (0,t_o)$. If derivatives $\frac{\partial\Phi_i}{\partial\xi}$ exist and are continuous near (0,0), then $\frac{\partial F}{\partial\eta}$ exists and is continuous near $(0,t_o)$. If q is even and $b > 0$, we may write

$$n - t_o = \pm\, \delta^{\lambda/q}[F(\delta,\eta)]^{\frac{1}{q}}.$$

Either branch may be solved for η in terms of δ by using the ordinary Implicit Function Theorem [ref. 11, p. 138], since the Jacobean is non-vanishing for small δ. If $b < 0$ there is no real solution. On the other hand if the multiplicity of t_o is odd, we may write

$$\eta - t_o = \delta^{\lambda/q}[F(\delta,\eta)]^{\frac{1}{q}}.$$

This one real branch can then be uniquely solved for any real b, again using the Implicit Function Theorem.

We now use the method of Dieudonné to prove the following result:

Theorem 4.2: Under H-1, H-2 and the supposition that (θ,λ_o) is an exceptional point of the nonlinear operator $T(x)$, (i.e., $\lambda_o I - T'(\theta)$ has no bounded inverse, or $\lambda_o \in P\sigma(T'(\theta))$), there exist two nontrivial solution branches $x^{\pm}(\lambda)$ of the equation $T(x) = \lambda x$ consisting of eigenfunctions of $T(x)$, which bifurcate from the trivial solution $x = \theta$ at the bifurcation point $\lambda = \lambda_o$. The two branches differ only in sign. If $a_2 a_4 < 0$, the two branches exist only for $\lambda > \lambda_o$ and bifurcation is said to be to the right; if $a_2 a_4 > 0$, the two branches exist only for $\lambda < \lambda_o$ and the bifurcation is said to be to the left. These branches exist at least in a small neighborhood of λ_o, and $\|x^{\pm}(\lambda)\| \to 0$ as $\lambda \to \lambda_o$.

Proof: We start with eq. (4.4). Clearly $\xi_1 = 0$ is a solution of eq. (4.4) for $\delta > 0$ or $\delta < 0$. Thus $u = \xi_1 u_1 = \theta$. Insertion of $u = \theta$ and $\pm\delta \neq 0$ in eq. (3.2) leads to the trivial solution.

Next, if we suppose $\xi_1 \neq 0$, it may be cancelled in eq. (4.4). There remains an equation in δ and ξ_1^2 which possesses a Newton Polygon with one side and slope -2. Assuming at first that $\delta > 0$, we put $\xi_1 = \eta\delta^{\frac{1}{2}}$. After cancelling δ, we get

$$[a_2+\Phi_2(\delta,\eta\delta^{\frac{1}{2}})] + \eta^2[a_4+\Phi_4(\delta,\eta\delta^{\frac{1}{2}})] = 0. \qquad (4.6)$$

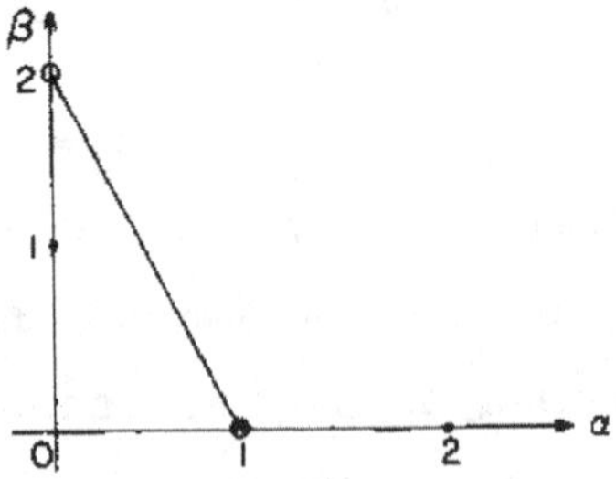

Newton Polygon

FIG. 4.2.

Solution of the leading part, $a_2 + a_4\eta^2 = 0$, leads to $\eta_{1,2} = \pm\sqrt{-\frac{a_2}{a_4}}$.

This represents two real solutions of unit multiplicity if and only if $a_2a_4 < 0$. Then with these roots we can rewrite eq. (4.6) as follows:

$$(\eta-\eta_1)(\eta-\eta_2) = -\Phi_2(\delta,\eta\delta^{\frac{1}{2}}) - \eta^2\Phi_4(\delta,\eta\delta^{\frac{1}{2}})$$
$$= \Lambda(\delta,\eta\delta^{\frac{1}{2}}) \to 0 \quad \text{as} \quad \delta \to 0,\ \delta > 0.$$

Since $\Lambda(\delta,\eta\delta^{\frac{1}{2}})$ is differentiable with respect to η, we can solve the two equations

$$\eta = \eta_1 + \frac{\Lambda(\delta,\eta\delta^{\frac{1}{2}})}{\eta-\eta_2} \qquad \eta = \eta_2 + \frac{\Lambda(\delta,\eta\delta^{\frac{1}{2}})}{\eta-\eta_1}$$

uniquely for η as a function of $\delta > 0$, employing the Implicit Function Theorem for real functions [ref. 11, p. 138], in a sufficiently small neighborhood. We get two real functions $\eta^{\pm}(\delta)$ for small $\delta > 0$, one tending to η_1 as $\delta \to 0$, the other to η_2. Through the relation $\xi_1 = \eta\delta^{\frac{1}{2}}$ there result two real curves $\xi_1^{\pm}(\delta)$ which, when substituted as ξ_1,δ pairs in eq. (3.2) with $u = \xi_1u_1$, provide two real solutions $x^{\pm}(\lambda)$ of $T(x) = \lambda x$ for λ near λ_0.

Clearly since $\xi_1 = \eta\delta^{\frac{1}{2}}$, $\delta > 0$, we see that $\xi_1^{\pm} \to 0$ as $\delta \to 0$ and thus $\|x^{\pm}(\lambda)\| \to 0$ as $\lambda \to \lambda_0$. Moreover because the use of the Implicit Function Theorem above implies a uniqueness property and because the Newton Polygon has

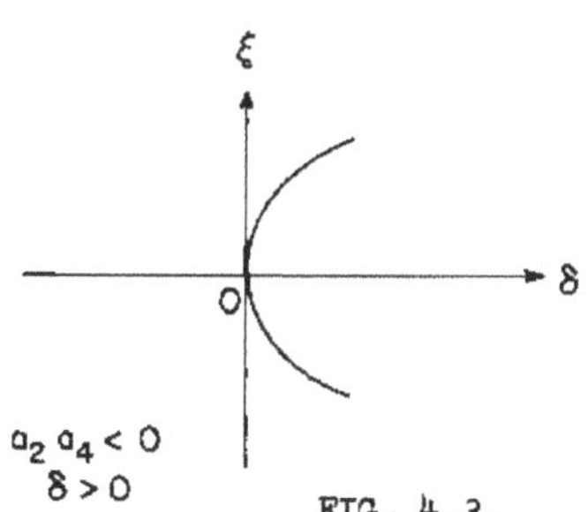

FIG. 4.3.

only one side, there are no other solutions of $T(x) = \lambda x$ such that $\|x\| \to 0$ as $\lambda \to \lambda_o$, $\lambda > \lambda_o$ for $a_2a_4 < 0$. By the oddness of $T(x)$, if $x(\lambda)$ is a solution of $T(x) = \lambda x$, so also is $-x(\lambda)$. Thus the two solution branches differ only in sign.

For $\delta < 0$, we substitute $\delta = -|\delta|$ into eq. (4.4). Then we put $\xi_1 = \eta|\delta|^{\frac{1}{2}}$ and cancel $\xi_1 \neq 0$ for nontrivial solutions. Solution of the leading part, $-a_2+a_4\eta^2 = 0$ now leads to $\eta_{1,2} = \pm\sqrt{\frac{a_2}{a_4}}$. There exist two real roots of unit multiplicity if and only if $a_2a_4 > 0$. The remainder of the analysis proceeds in exactly the same way as with the case $\delta > 0$.

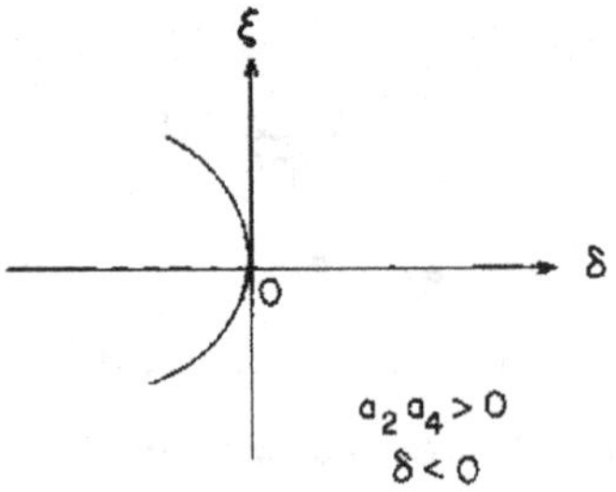

FIG. 4.4.

We have two mutually exhaustive situations represented here. If $a_2a_4 < 0$ we have produced exactly two real nontrivial solutions for $\delta > 0$, while for $\delta = -|\delta| < 0$ we have seen there are no real solutions. Likewise if $a_2a_4 > 0$ we have seen that there are no real solutions for $\delta > 0$ (the leading part in eq. (4.6), namely $a_2 + a_4\eta^2 = 0$ has no real roots), while for $\delta = -|\delta| < 0$ we have produced exactly two real nontrivial solutions. This ends the proof of Theorem 4.2.

Of course the hypotheses of Theorem 4.2 are unnecessarily stringent. Since the theorem furnishes branch solutions of the equation $T(x) = \lambda x$

only near $\lambda = \lambda_o$, and since the bifurcation equation, eq. (4.1) is valid only with a restriction on ξ_1: $|\xi_1| \le e$, where e is a constant in Theorem 3.2, we see that Theorem 4.2 is really only a local theorem. In its statement we need only assume oddness of $T(x)$ in a neighborhood of the origin $x = \theta$.

In writing eq. (4.4), part of the assumption of oddness was that $d^2T(\theta;h_1,h_2) = \theta$, $h_1,h_2 \in X$. This leads us to our next theorem, which we include for completeness:

Theorem 4.3: Under H-1 but not H-2, and with the supposition that $d^2T(\theta;h_1,h_2) \neq \theta$, the two branches of eigenfunctions of the operator $T(x)$ $(T(\theta) = \theta)$ which bifurcate from the trivial solution $x_o = \theta$ at the bifurcation point $\lambda = \lambda_o$, exist, one on each side of λ_o. These branches exist at least in a small neighborhood of $\lambda = \lambda_o$, and $\|x(\lambda)\| \to 0$ as $\lambda \to \lambda_o$, $\lambda \gtrless \lambda_o$.

Proof: In this case, eq. (4.1) is written as follows:

$$-\delta\xi_1 u_1^* u_1 + \frac{\xi_1^{\,2}}{2} u_1^* d^2T(\theta;u_1u_1) + \frac{\xi_1^{\,3}}{6} u_1^* d^3T(\theta;u_1,u_1,u_1) + \sum_{i+j=1}^{3} u_1^*\omega_{ij}(\delta,u_1) = 0.$$

Since we can show again that $\omega_{10} = \omega_{20} = \omega_{30} = 0$, the bifurcation equation may be put in the form (compare with eq. 4.2):

$$\xi_1\delta[a_2+\Phi_2(\delta,\xi_1)] + \xi_1^{\,2}[a_3+\Phi_3(\delta,\xi_1)] + \xi_1^{\,3}[a_4+\Phi_4(\delta,\xi_1)] = 0,$$

with a_2, a_3, a_4 and $\Phi_2(\delta,\xi_1)$, $\Phi_3(\delta,\xi_1)$, $\Phi_4(\delta,\xi_1)$ as defined in connection with eq. (4.2). By putting $\overline{\Phi_3}(\delta,\xi_1) = \Phi_3(\delta,\xi_1) + \xi_1[a_4+\Phi_4(\delta,\xi_1)]$, we may also write this bifurcation equation in the form, [ref. 8, p. 50]:

$$\xi_1\delta[a_2+\Phi_2(\delta,\xi_1)] + \xi_1^2[a_3+\overline{\Phi_3}(\delta,\xi_1)] = 0. \tag{4.7}$$

After cancellation of $\xi_1 \neq 0$ ($\xi_1 = 0$ leads to the trivial solution), eq. (4.7) is an equation in δ, ξ_1 with a one-sided Newton Polygon with slope -1. Putting $\xi_1 = \eta\delta$ we have, for $\delta \neq 0$,

$$[a_2+\Phi_2(\delta,\eta\delta)] + \eta[a_3+\overline{\Phi_3}(\delta,\eta\delta)] = 0. \tag{4.8}$$

The leading part, $a_2 + a_3\eta = 0$, has the single root, $\eta_0 = -\frac{a_2}{a_3}$, regardless of the sign of a_2a_3. Then eq. (4.8) is put into the form

$$\eta-\eta_0 = -\Phi_2(\delta,\eta\delta) - \eta\overline{\Phi_3}(\delta,\eta\delta)$$
$$= \Lambda(\delta,\eta\delta) \to 0 \text{ as } \delta \to 0. \tag{4.9}$$

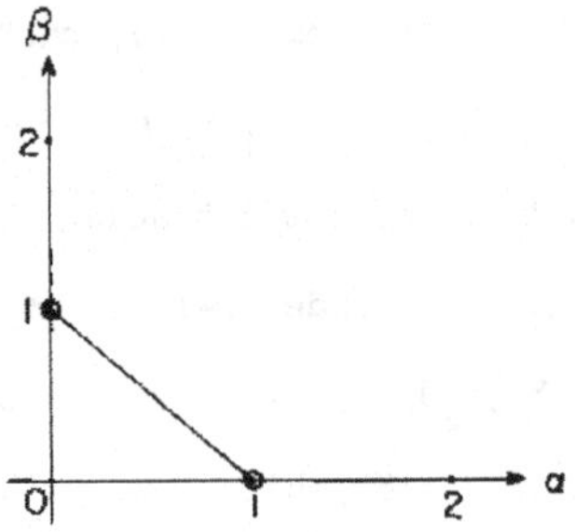

Newton Polygon

FIG. 4.5.

Since $\Lambda(\delta,\eta\delta)$ is differentiable with respect to η in a neighborhood of $\eta = \eta_0$, we can employ the Implicit Function Theorem for real functions [ref. 11, p. 138] to produce a solution $\eta(\delta)$ of eq. (4.9) whether $\delta > 0$ or $\delta < 0$. Through the relationship $\xi_1 = \eta\delta$, we have a unique real function $\xi_1(\delta)$ for $\delta \gtrless 0$ which when substituted as ξ_1, δ

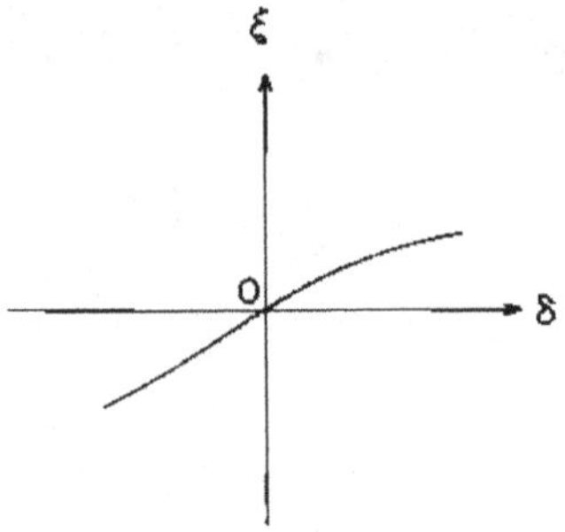

FIG. 4.6.

pairs in eq. (3.2) provide unique small solutions $x(\lambda)$ of $T(x) = \lambda x$ for λ near λ_o. Moreover since $\xi_1(\delta) \to 0$ as $\delta \to 0$, we see by means of eq. (3.2) that $\|x(\lambda)\| \to 0$ as $\lambda \to \lambda_o$. This ends the proof.

To end the present section of these notes we present two rather simple examples which might be somewhat illustrative of the foregoing methods.

Example: Let us solve the integral equation

$$\lambda\varphi(s) = \int_0^{2\pi} \sin s \sin t \, [\varphi(t)+\varphi^2(t)]dt$$

of quadratic type with first rank kernel.

Eq. (2.3) is represented in this case as follows:

$$\left[\lambda_o I - \int_0^{2\pi} \sin s \sin t \, [1+2\phi_o]\cdot dt\right]h = -\delta\phi_o - \delta h + \int_0^{2\pi} \sin s \sin t \; h^2(t)dt = F_\delta(h)$$

where we assume that λ_o, ϕ_o is a solution of the problem. There is no remainder term.

The linearized problem

$$\lambda_o h - \int_0^{2\pi} \sin s \sin t[1+2\phi_o(t)]h(t)dt = 0$$

has an eigenvalue $\lambda_o = \int_0^{2\pi} \sin^2 t[1+2\phi_o(t)]dt$ and the normalized eigenfunction $\sin s$. Since the kernel is symmetric, λ_o has Riesz index unity [ref. 21, p. 342, Th. 18], where we assume that our Banach space is $L_2(0,2\pi)$. The null space η of the operator $\lambda_o I - \int_0^{2\pi} \sin s \sin t[1+2\phi_o]\cdot dt$ is one dimensional. Let E project on η; then since the Riesz index is unity, $I-E$ projects on the range R, [ref. 19, p. 183, p. 217].

Let us first treat bifurcation at the origin, i.e., let $\phi_0 \equiv 0$. Eq. (3.2) is represented as follows:

$$h = \widetilde{M}F_\delta(h) + \xi \sin s = M(I-E)\left\{ -\delta h + \int_0^{2\pi} \sin s \sin t \, h^2 \, dt \right\} + \xi \sin s$$

$$= M\{-\delta(I-E)h+0\} + \xi \sin s$$

where ξ is a scalar. The iterations indicated in Theorem 3.1 are trivial, so that the solution is $h = V_\delta(u) = \xi \sin s$. When this is substituted in eq. (3.4), we have

$$EF_\delta(V_\delta(u)) = EF_\delta(\xi \sin s) = E\left\{ -\delta\xi \sin s + \xi^2 \int_0^{2\pi} \sin s \sin^3 t \, dt \right\}$$

$$= E\{-\delta\xi \sin s\} = -\delta\xi \sin s = 0.$$

As solutions of the bifurcation equation, either $\delta \neq 0$, $\xi = 0$ or $\delta = 0$, $\xi \neq 0$ are possible. The former is the trivial solution, which leads to $h \equiv 0$. The latter is nontrivial and leads to a branch of eigensolutions of the nonlinear problem in a neighborhood of $\phi_0 \equiv 0$, namely $\varphi = \xi \sin s$, where ξ is a completely arbitrary scalar. This branch of eigenfunctions exists only for the single value $\lambda = \lambda_0$.

Thus in this case, a nonlinear operator possesses an eigenvalue λ_0, and an associated eigenspace: that space which is spanned by sin s. In this respect it is like a linear problem.

In this example the bifurcation equation at the origin yields solutions valid in the large because the iteration process for eq. (3.2) is finite in duration, and the local requirements of Theorem 3.2 are not necessary.

Example: Now let us deal similarly with the cubic equation:

$$\lambda\varphi(s) = \int_0^{2\pi} \sin s \sin t \, [\varphi(t)+\varphi^3(t)]dt$$

for which the linearization at the origin again has the single eigenvalue $\lambda_0 = \int_0^{2\pi} \sin^2 t \, dt = \pi$ of unit Riesz index, and the eigenfunction sin s. Again let E project on the one dimensional null space η spanned by sin s. Then with $\varphi_0 \equiv 0$, we have eq. (3.2) represented as follows:

$$h = \overline{M}F_\delta(h) + \xi \sin s = M(I-E)\left\{-\delta h + \int_0^{2\pi} \sin s \sin t \, h^3(t)dt\right\} + \xi \sin s$$

$$= M\{-\delta(I-E)h+0\} + \xi \sin s.$$

Again the iteration process is trivial, terminating with the solution $h = V_\delta(u) = \xi \sin s$. The bifurcation equation (eq. 3.4)) becomes

$$EF_\delta(\xi \sin s) = E\left\{-\delta\xi \sin s + \xi^3 \int_0^{2\pi} \sin s \sin^4 t \, dt\right\} = \left\{-\delta\xi + \frac{3\pi}{4}\xi^3\right\} \sin s = 0$$

whence either $\xi = 0$, δ arbitrary, or $-\delta + \frac{3\pi}{4}\xi^2 = 0$. The former possibility leads to the trivial solution $\phi \equiv 0$; the latter case is that of a right-facing parabolic curve. We are lead to a branch of nontrivial eigenfunctions parametrized by λ: $\varphi = \pm\sqrt{\frac{\lambda-\lambda_0}{3\pi/4}} \sin s$. Thus the eigenfunctions in this example do not form a linear space as in linear problems and the previous nonlinear example. Rather they form a nonlinear manifold, which has been given the name: "continuous branch."

Again, since the iterations are trivial, the sort of local restrictions imposed in Theorem 3.2 do not apply, and we have produced here solutions in the large. Are they the only solutions?

Further solutions could be produced if there were bifurcation on the continuous branch at some point other than the origin. At $\lambda_1 > \lambda_o = \pi$, the continuous branch yields the eigenfunction $\varphi_1 = \sqrt{\frac{\lambda_1-\lambda_o}{3\pi/4}}\sin s$; if this is a bifurcation point, the operator $\lambda_1 I - \int_0^{2\pi} \sin s \sin t \left[1 + \frac{4}{\pi}(\lambda_1-\lambda_o)\sin^2 t\right] \cdot dt$ is singular. It can be seen however that the linearized problem at λ_1, $\sqrt{\frac{\lambda_1-\lambda_o}{3\pi/4}}\sin s$, namely

$$\lambda h - \int_0^{2\pi} \sin s \sin t \left[1 + \frac{4}{\pi}(\lambda_1-\lambda_o)\sin^2 t\right] h(t)dt = 0,$$

has $\lambda = \lambda_1 + 2(\lambda_1-\lambda_o) > \lambda_1$ as its only eigenvalue, corresponding to eigenfunction sin s. Since λ_1,φ_1 could be any point on the continuous branch, we see that a "secondary bifurcation" does not exist in this problem. There is no point of the branch where the considerations of Theorem 3.3 are applicable.

This example is illustrative of the situation of Theorem 4.2. A similar problem illustrative of Theorem 4.3 would be

$$\lambda\varphi(s) = \int_0^{\pi} \sin s \sin t[\varphi(t)+\varphi^2(t)+\varphi^3(t)]dt.$$

5. The Eigenvalue Problem; Hammerstein Operators; Sublinear and Superlinear Operators; Oscillation Kernels.

In the preceding development we have discussed the extension of branches of eigenelements of quite general nonlinear operators $T(x)$ of a Banach Space X into itself. Then we treated bifurcation of branches of eigenelements of these general operators under the assumption that $T'(x_0)$ is compact at a bifurcation point (x_0,λ_0) (condition H-1); this can be accomplished if $T(x)$ itself is assumed to be completely continuous, in which case $T'(x)$ is compact everywhere. There resulted a set of simultaneous bifurcation equations of quite a general character, namely eq. (3.8). Because of the difficulties in handling the general bifurcation equations, attention was then confined to the case where the null space $\eta_1(x_0)$ of the operator $\lambda_0 I-T'(x_0)$ is one dimensional, where (x_0,λ_0) is the bifurcation point. In this case eq. (3.8) becomes just one scalar equation in one scalar unknown. The Newton Polygon method was used to treat this case of "bifurcation at an eigenvalue of $T'(x_0)$ of unit multiplicity." The treatment became very explicit in the case of an odd operator: $T(-x) = - T(x)$, in which case $T(\theta) = \theta$, where $\theta \in X$ is the null element. We handled "bifurcation at the origin" in Theorem 4.2; i.e. we set $x_0 = \theta$.

With odd operators $T(x)$ and bifurcation at the origin $x_0 = \theta$, we have a situation which may roughly be compared with the situation pertaining to compact linear operators on a real Banach space X, and the real eigenvalue problem for such operators.

The eigenvalue problem for an odd completely continuous operator $T(x)$ may be described as follows: Find those values of the real parameter λ such that the equation

$$\lambda x = T(x) \tag{5.1}$$

has nontrivial solutions $x \in X$, and then explore the properties of these nontrivial solutions. Eq. (5.1) always has of course the trivial solution by the oddness of $T(x)$. Now if $T(x) = Ax$, $x \in X$, where A is a completely continuous linear operator, the study of eq. (5.1) leads along familiar paths. The eigenvalues form a discrete sequence $\left\{\mu_n^{(o)}\right\}$ of real numbers; these eigenvalues are of finite algebraic multiplicity, and thus of finite geometric multiplicity (ref. 21, p. 336). Associated with each eigenvalue therefore is a finite dimensional linear space, sometimes called an eigenspace; this space is spanned by the eigenelements associated with the eigenvalue.

Since the eigenspace is a linear space, it contains elements of arbitrarily large norm. If we were to make a two-dimensional plot of the norm of the eigenelement vs. the eigenvalue, we should have an array of vertical lines emanating from the λ axis and extending to infinity. For linear operators, such a portrayal would not seem to have any conceptual advantages.

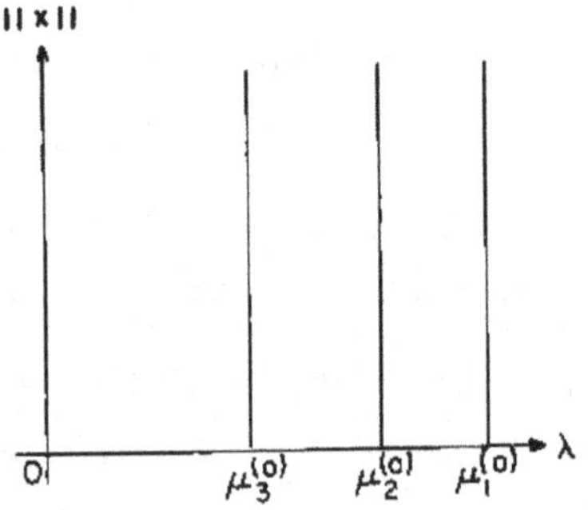

FIG. 5.1.

If we think of a linear operator, however, as a type of nonlinear operator, we can regard these vertical lines as representing branches of eigenelements and the eigenvalues $\left\{\mu_n^{(o)}\right\}$ as being bifurcation points at the origin. If the eigenvalues $\left\{\mu_n^{(o)}\right\}$ are of unit multiplicity, this description is an apt one. Indeed, there are nonlinear operators with linear eigenspaces bifurcating from the trivial solution $x = \theta$, as exemplified by the first example at the end of Section 4.

In general with nonlinear odd operators, the eigenelements $x(\lambda)$ are not such trivial functions of the eigenvalues λ, and the branches of eigenelements which bifurcate from an eigenvalue $\mu_n^{(o)}$ of the linearized operator $T'(\theta)$ are nonlinear manifolds. On a norm vs. λ plot we do not in general have vertical straight lines. The questions asked though are the same, i.e. those asked in connection with eq. (5.1).

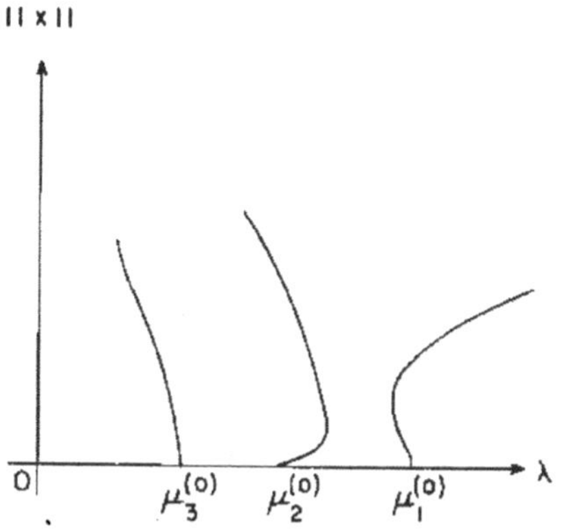

FIG. 5.2.

The problem therefore in dealing with eq. (5.1), for an odd completely continuous operator $T(x)$, is to take the information developed in Theorem 4.2 about the "primary bifurcation points" $\left\{\mu_n^{(o)}\right\}$, (i.e. $\lambda_o = \mu_n^{(o)}$ in that theorem), and about the corresponding solution branches in a small neighborhood of $x = \theta$, and then to extend these branches into the large. In other words, it usually is not enough to know the behavior of the branches of eigenelements merely in a small neighborhood of the origin in real Banach space X. We want to study eigenelements with large norms also.

When it comes to this question of extending branches of eigensolutions from the small to the large, we run out of general theory. Branches can be extended by stepwise application of the process of Theorem 2.3 provided there is a pair (x,λ) on a branch which is recognizable as an "ordinary point," i.e., where $\lambda I-T'(x)$ has a bounded inverse. The considerations of Theorem 2.4 are applicable, however, which means we cannot get past the "exceptional points" which may occur, i.e., pairs (x,λ) where $\lambda I-T'(x)$ has no inverse. In order to investigate this latter problem we must now considerably restrict the class of operators $T(x)$. Namely we consider the operators of Hammerstein, (ref. 14, p. 46).

Generally, a Hammerstein operator consists of an operator of Nemytskii [ref. 14, p. 20], namely $fx = f(s,x(s))$, defined on some function space, premultiplied by a linear operator K, i.e. we let $T(x) = Kfx$.

In the sequel, Hammerstein operators will play the leading role. We admit at this point that we are not partial to Hammerstein operators on account of any great applicability in physical problems, though there are a few such applications, of course. One might mention the rotating chain problem [ref. 38] and the rotating rod problem [refs. 2, 20]. Rather, we like Hammerstein operators because they are amenable to the study of branches of eigenfunctions in the large. Assumptions can be made about the linear operator K which make the study presently possible. An extension of these results to other classes of nonlinear operators is to be desired, but seems difficult now.

In the present study, we let the Banach space X be the space $C(0,1)$ of real continuous functions $x(s)$ defined on the interval $0 \leq s \leq 1$ with the sup norm. The Nemytskii operator f is defined by a function f(s,x) which is continuous in the strip $0 \le s \le 1$, $-\infty < x < +\infty$, uniformly in x with respect to s. The linear operator K is generated by a bounded continuous kernel K(s,t), and thus is compact on C(0,1). Thus we let

$$T(x) = Kfx = \int_0^1 K(s,t)f(t,x(t))dt. \tag{5.2}$$

The discrete Hammerstein operator is defined on finite dimensional vector space by letting K be a square matrix and f a nonlinear vector function. It is interesting for examples, but is not fundamentally different than the operator of eq. (5.2).

We further assume that f(s,x) is differentiable with respect to x, uniformly in s, up to the third order, and we define the following Fréchet derivatives:

$$T'(x)h = Kf_x'h = \int_0^1 K(s,t)f_x'(t,x(t))h(t)dt,$$

$$T''(x)h_1h_2 = Kf_x''h_1h_2 = \int_0^1 K(s,t)f_x''(t,x(t))h_1(t)h_2(t)dt,$$

$$T'''(x)h_1h_2h_3 = Kf_x'''h_1h_2h_3 = \int_0^1 K(s,t)f_x'''(t,x(t))h_1(t)h_2(t)h_3(t)dt$$

which are defined everywhere on C(0,1) since C(0,1) is an algebra.

All of the theory developed in sections 2-4 holds for these operators.

It is very useful in the study of branches of eigenfunctions in the large to distinguish two types of nonlinearity. We assume we are speaking of odd operators: $T(-x) = -T(x)$, which for Hammerstein operators means that $f(s,-x) = - f(s,x)$. Thus $f(s,0) \equiv 0$. We further assume that $f_x'(s,x) > 0$, $0 \le s \le 1$, $-\infty < x < +\infty$. Then for Hammerstein operators we readily distinguish the following pure categories of nonlinearity:

(1) Sublinearity:

$$xf_x''(s,x) < 0$$

$$0 \le s \le 1, \; -\infty < x < +\infty$$

from which it follows that

$$f_x'''(s,x) < 0;$$

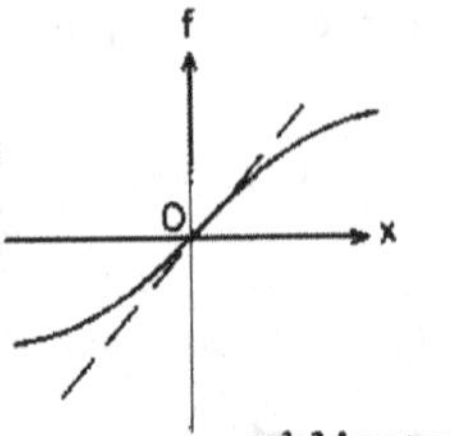

sublinear

FIG. 5.3a.

(2) Superlinearity:

$$xf_x''(s,x) > 0$$

$$0 \le s \le 1, \; -\infty < x < +\infty$$

from which it follows that

$$f_x'''(s,x) > 0.$$

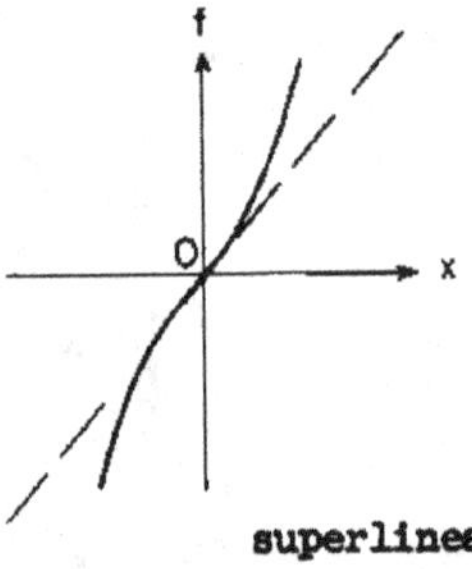

superlinear

FIG. 5.3b.

A Hammerstein operator Kf, with such conditions on f, is said to be respectively a sublinear or a superlinear Hammerstein operator.

Obviously these two types of nonlinearity are not exhaustive; generally a nonlinear operator could be of neither type. In physical applications, however, problems seem to be preponderantly of pure sublinear or superlinear type, and these qualities certainly transcend the Hammerstein class of operators.

There is another important type of operator, where our classification goes in a somewhat different direction: the Asymptotically Linear operator. A Hammerstein operator is asymptotically linear if

$$\lim_{x\to\pm\infty} \frac{f(s,x)}{x} = A(s) \neq 0.$$

An asymptotically linear Hammerstein operator can obviously be sublinear, superlinear, or neither. If it is either sublinear or superlinear, however, then

$$\lim_{x\to\pm\infty} \frac{f(s,x)}{x} = A(s) \geq 0.$$

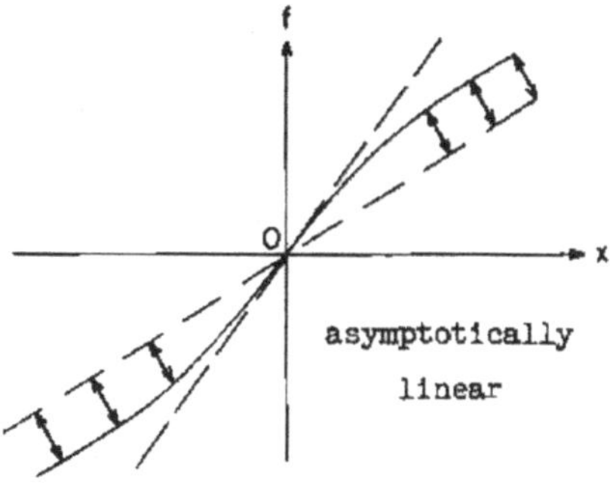

FIG. 5.4.

The example of section 1 is a superlinear problem, but not asymptotically linear.

Let us see how our bifurcation theory appears with Hammerstein operators, and one or the other assumption of sublinearity or superlinearity. Suppose λ_0 is an eigenvalue of Kf_x' for $x = x_0(s)$, where $x_0(s)$ satisfies Hammerstein's equation $Kfx = \lambda x$ for some λ, of multiplicity unity. Suppose that u_1 is the associated normalized eigenfunction, i.e., $Kf_x'u_1 = \lambda_0 u_1$, and $u_1{*}$ the eigenfunction of the adjoint problem: $f_x'K^*u_1{*} = f_x'(s,x_0(s)) \int_0^1 K(t,s)u_1{*}(t)dt = \lambda_0 u_1{*}(t)$. With reference to the explicit bifurcation equation, namely eq. (4.2), we put down the coefficients of that equation in the Hammerstein case (here we use the setwise imbedding $C(0,1) \subset L_2(0,1)$, and employ the inner product):

$$a_1 = -(u_1^*, x_o)$$

$$a_2 = -\left(u_1^*, u_1 + \int_0^1 K(s,t) f_x''(t, x_o(t)) u_1(t) \bar{M} x_o(t) dt\right)$$

$$a_3 = \frac{1}{2}\left(u_1^*, \int_0^1 K(s,t) f_x''(t, x_o(t)) u_1^2(t) dt\right)$$

$$a_4 = \frac{1}{6}\left(u_1^*, \int_0^1 K(s,t) f_x'''(t, x_o(t)) u_1^3(t) dt\right).$$

It is very convenient to assume now that $K(s,t) = K(t,s)$; i.e. $K(s,t)$ is symmetric. Then $u_1^*(s) = f_x'(s, x_o(s)) u_1(s)$, and

$$\begin{aligned} a_1 &= -(f_x'(s, x_o(s)) u_1, \ x_o(s)) = -(f'_{x_o} u_1, x_o) \\ a_2 &= -(f'_{x_o} u_1, u_1) - \lambda_o (u_1, f''_{x_o} u_1 \bar{M} x_o) \\ a_3 &= \frac{1}{2} \lambda_o (u_1, f''_{x_o} u_1^2) = \frac{1}{2} \lambda_o (f''_{x_o}, u_1^3) \\ a_4 &= \frac{1}{6} \lambda_o (u_1, f'''_{x_o} u_1^3) = \frac{1}{6} \lambda_o (f'''_{x_o}, u_1^4). \end{aligned} \tag{5.3}$$

Taking $x_o \equiv 0$ so as to study bifurcation at the origin, we have for these coefficients:

$$a_1 = 0, \ a_2 = -(f_x'(s,0), u_1^2), \ a_3 = \frac{\lambda_o}{2} (f_x''(s,0), u_1^3), \ a_4 = \frac{\lambda_o}{6} (f_x'''(s,0), u_1^4)$$

where of course $u_1(s)$ is the normalized solution of the problem $\lambda_o h(s) = \int_0^1 K(s,t) f_x'(s,0) h(s) ds$.

Let us examine the sublinear and superlinear cases and the signs which the above coefficients assume. In the sublinear case, we have for $\lambda_0 > 0$ that $a_1 = a_3 = 0$, $a_2 < 0$ and $a_4 < 0$ so that $a_2a_4 > 0$. With reference to eq. (4.4) and Theorem 4.2, there are small real solutions for $\lambda < \lambda_0$, none if $\lambda > \lambda_0$. Thus bifurcation is to the left at $\lambda = \lambda_0$. Again if $\lambda_0 < 0$, we have $a_2a_4 < 0$ and bifurcation is to the right.

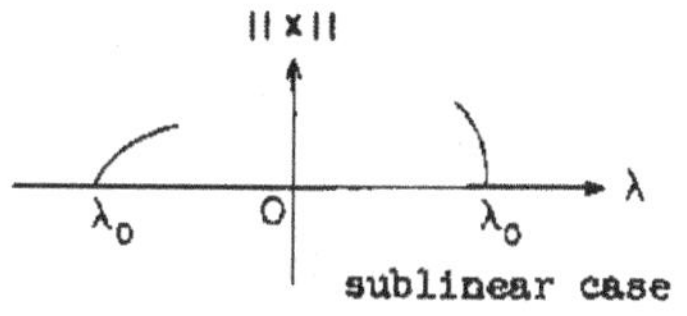

FIG. 5.5a.

In the superlinear case, if $\lambda_0 > 0$, we have $a_1 = a_3 = 0$, $a_2 < 0$ and $a_4 > 0$ so that $a_2a_4 < 0$. With reference to Theorem 4.2, there exist small real solutions for $\lambda > \lambda_0$, none for $\lambda < \lambda_0$. Hence bifurcation is to the right at $\lambda = \lambda_o$. If $\lambda_0 < 0$, we have $a_2a_4 > 0$ so that bifurcation is to the left.

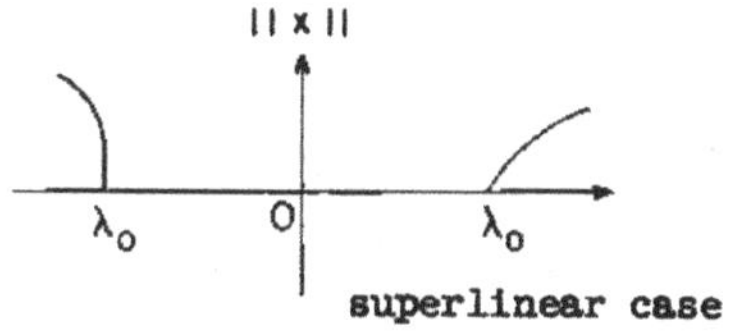

FIG. 5.5b.

The above remarks on sublinearity and superlinearity have an analog with abstract operators. Indeed let $X = H$, a real Hilbert space, and let $T(x)$ be an odd operator: $T(-x) = -T(x)$, with $T(\theta) = \theta$. Further we suppose that $T(x)$ is completely continuous and of variational type [ref. 14, p. 300]; in this case $T'(x)$ is compact and symmetric for given $x \in H$. Suppose moreover that $T'(x)$ is positive definite for given $x \in H$.

Such an operator $T(x)$ is said to be sublinear if $(dT'(x,x)h,h) < 0$ for all h, $x \in H$. In other words $dT'(x;x) = T''(x)x$ is, for all $x \in H$, a negative definite linear transformation of H into itself. Similarly $T(x)$ is said to be superlinear if $(dT'(x,x)h,h) > 0$ for all h, $x \in X$.

With $\alpha > 0$ any number, and $x \in H$, we have by definition of the Fréchet differential [ref. 15, p. 183] and the fact that $T''(\theta)x = \theta$,

$$\begin{aligned} dT'(\alpha x;x) &= T''(\alpha x)x = T''(\alpha x)x - T''(\theta)x \\ &= dT''(\theta;\alpha x)x + R(\theta,\alpha x)x, \end{aligned} \tag{5.4}$$

where $R(\theta,\alpha x) = o(\alpha\|x\|)$. Since in the <u>sublinear</u> case $(dT'(\alpha x;x)h,h) < 0$ for all h, $x \in H$, it can be seen from eq. (5.4) that for α small enough, $(dT''(\theta;\alpha x)xh,h) = (d^2T'(\theta;\alpha x,x)h,h) < 0$; this implies however that $(d^2T'(\theta;x,x)h,h) < 0$ for all h, $x \in H$. Similarly, for the <u>superlinear case</u> $(d^2T'(\theta;x,x)h,h) > 0$ for all h, $x \in H$.

Then for $x_0 = \theta$, we have the following coefficients in the bifurcation equation, eq. (4.2), for the sublinear case:

$a_1 = 0$, $a_2 = -(u_1,u_1) = -1$, $a_3 = 0$

$$a_4 = \frac{1}{6}(u_1,d^3T(\theta;u_1,u_1,u_1)) = \frac{1}{6}(u_1,d^2T'(\theta;u_1,u_1)u_1) < 0.$$

Here of course, $(\lambda_0 I - T'(\theta))u_1 = \theta$. Thus in the sublinear case, $a_2a_4 > 0$ and we have bifurcation to the left. On the other hand, if it were the superlinear case, we should have $a_2a_4 < 0$ and bifurcation to the right, (see Theorem 4.2).

Perhaps the chief reason for our selection of Hammerstein operators as an object of study is the fact that this type of concrete nonlinear operator possesses a separated kernel $K(s,t)$ about which we can make further assumptions. Specifically, from an investigative standpoint, it is useful to assume that $K(s,t)$ is an oscillation kernel, [ref. 9, p. 236].

Definition: An $n \times n$ matrix $A = (a_{ik})$ is a completely non-negative matrix (or respectively completely positive) if all its minors of any order are non-negative (or respectively positive).

Definition: An $n \times n$ matrix $A = (a_{ik})$ is an oscillation matrix if it is a completely non-negative matrix, and there exists a positive integer κ such that A^{κ} is a completely positive matrix.

Definition: A continuous kernel $K(s,t)$, $0 \le s, t \le 1$, is an oscillation kernel if for any set of n points $x_1, x_2, \ldots, x_n$, where $0 \le x_i \le 1$, one of which is internal, the matrix $(K(x_i,x_k))_1^n$ is an oscillation matrix, $n = 1, 2, 3, \ldots$.

With $K(s,t)$ a symmetric oscillation kernel, we have the following properties for eigenvalues and eigenfunctions of the equation

$$\lambda\phi(s) = \int_0^1 K(s,t)\phi(t)d\sigma(t) \tag{5.5}$$

where $\sigma(t)$ is a non-diminishing function with at least one point of growth in the open interval $0 < t < 1$, [ref. 9, p. 262]:

(a) There is an infinite set of eigenvalues if $\sigma(t)$ has an infinite number of growth points.

(b) All the eigenvalues are positive and simple: $0 < \ldots < \lambda_n < \lambda_{n-1} < \ldots < \lambda_o$.

(c) The eigenfunction $\phi_o(s)$ corresponding to λ_o has no zeros on the open interval $0 < s < 1$.

(d) For each $j = 1, 2, \cdots$, the eigenfunction $\phi_j(s)$ corresponding to λ_j has exactly j nodes (odd order zeros) in the interval $0 < s < 1$, and no other zeros.

(e) $\phi(s) = \sum_{i=k}^{m} c_i \phi_i(s)$ has at most m zeros and at least k nodes in the interval $0 < t < 1$, for given c_i, $\sum_{i=k}^{m} c_i^2 > 0$. If the number of zeros is equal to m, these zeros are nodes.

(f) The nodes of the functions $\phi_j(s)$ and $\phi_{j+1}(s)$ alternate, $j = 1, 2, \ldots$.

Our interest in the oscillation kernel in dealing with Hammerstein operators stems from the fact that with $f_x'(s,x) > 0$, $0 \le s \le 1$, $-\infty < x < +\infty$ as we have supposed, the Fréchet derivative

$$Kf_x'h = \int_0^1 K(s,t)f_x'(t,x(t))h(t)dt, \tag{5.6}$$

with $K(s,t)$ an oscillation kernel, is a case of a linear operator such as that appearing in eq. (5.5), so that the properties (a)-(f) listed above are true for its eigenvalues and eigenfunctions. We wish to stress as very important for Hammerstein operators that properties (a)-(f) hold for operator (5.6) whatever the continuous function $x(t)$ used in the definition of the operator, if $K(s,t)$ is an oscillation kernel.

Properties (e),(f) are actually in excess of requirement as far as we know, as is also the statement in property (b) about the positivity of the eigenvalues.

With $K(s,t)$ an oscillation kernel, every eigenvalue $\mu_p^{(o)}$, $p = 0, 1, 2, \ldots$ of the Fréchet derivative $Kf_o'h = \int_0^1 K(s,t)f_x'(t,o)h(t)dt$ at the origin is of multiplicity unity, so that Theorem 4.2 or 4.3 is directly applicable to study primary bifurcation from the trivial solution. Each such eigenvalue $\mu_p^{(o)}$ is a bifurcation point. Moreover if $x_0(s,\lambda_0)$ is an exceptional point on a branch of eigensolutions, i.e. the Fréchet derivative $Kf_{x_0}'h = \int_0^1 K(s,t)f_x'(t,x_0(t))h(t)dt$ has an eigenvalue λ_0, or $\lambda_0 I - Kf_{x_0}'$ has no bounded inverse, then we know a priori that λ_0 is a simple eigenvalue, or the null space $\eta_1(x_0)$ is one dimensional. Hence our bifurcation theory with the Newton Polygon method is applicable, in particular eq. (4.2).

Another benefit in assuming an oscillation kernel is illustrated in the following example for a discretized Hammerstein operator:

Example: Consider the discrete superlinear problem:

$$\begin{pmatrix} a & 0 \\ 0 & b \end{pmatrix} \begin{pmatrix} u + u^3 \\ v + v^3 \end{pmatrix} = \lambda \begin{pmatrix} u \\ v \end{pmatrix}, \qquad a > b \tag{5.7}$$

for which we have the following linearization:

$$\begin{pmatrix} a & 0 \\ 0 & b \end{pmatrix} \begin{pmatrix} (1+3u^2)h \\ (1+3v^2)k \end{pmatrix} = \lambda \begin{pmatrix} h \\ k \end{pmatrix}. \tag{5.8}$$

At the origin, $u = v = 0$, and we have primary bifurcation points $a > b$. A continuous branch of eigenvectors, namely $\begin{pmatrix} \pm\sqrt{\frac{\lambda}{a} - 1} \\ 0 \end{pmatrix}$ bifurcates to the right at $\lambda = a$ from the trivial solution $\begin{pmatrix} 0 \\ 0 \end{pmatrix}$ while another branch $\begin{pmatrix} 0 \\ \pm\sqrt{\frac{\lambda}{b} - 1} \end{pmatrix}$

bifurcates to the right at $\lambda = b$. Of interest is the behavior of the eigenvalues of the linearized problem eq. (5.8) as the branches evolve. Taking the second branch $\begin{pmatrix} 0 \\ \pm\sqrt{\frac{\lambda}{b} - 1} \end{pmatrix}$, and letting $u = 0$ and $v = \pm\sqrt{\frac{\lambda}{b} - 1}$ in eq. (5.8)

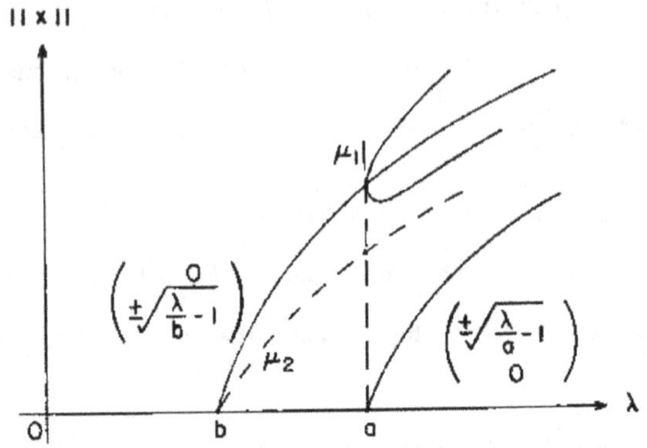

FIG. 5.6.

we see that the linearization has two eigenvalues, $\mu_1 = a$ and $\mu_2 = 3\lambda - 2b$. The parameter λ increases as the second branch evolves however, and whereas initially we have $\mu_2 < \mu_1$, for $\lambda > \frac{a+2b}{3}$ we have $\mu_2 > \mu_1$. Moreover a situation is attained where $\lambda = \mu_1 = a$. At this point on the second branch, eq. (5.8) has the nontrivial solution $\begin{pmatrix} 1 \\ 0 \end{pmatrix}$ with $\lambda = \mu_1 = a$, and a secondary bifurcation takes place.

In this example, the kernel or matrix $\begin{pmatrix} a & 0 \\ 0 & b \end{pmatrix}$ is not an oscillation matrix; two of its minors vanish. If we were to have used an oscillation matrix, say $\begin{pmatrix} a & \epsilon \\ \epsilon & b \end{pmatrix}$ with $0 < \epsilon < \sqrt{ab}$, we could have been assured a priori that eigenvalues μ_1, μ_2 would always be simple, and the crossover point at $\lambda = \frac{a+2b}{3}$ would have been impossible. We should have had $\mu_2 < \mu_1$ always, and no secondary bifurcation.

Green's functions for differential operators are often oscillation kernels [ref. 9, p.312]. Therefore when Hammerstein operators result from boundary value problems for differential equations in applied work, a Hammerstein equation involving an oscillation kernel must often be solved.

6. On the Extension of Branches of Eigenfunctions; Conditions Preventing Secondary Bifurcation of Branches.

In this section we assume that $T(x) = \int_0^1 K(s,t)f(t,x(t))dt$, i.e., $T(x)$ is a Hammerstein operator. The kernel $K(s,t)$ is bounded and continuous, $0 \le s \le 1$, $0 \le t \le 1$. The function $f(s,x)$ is continuous in the strip $0 \le s \le 1$, $-\infty < x < +\infty$ uniformly in x with respect to s.

Thus if $f_x'(s,x)$ exists and is integrable, $T(x)$ satisfies H-1 as it stands; i.e., $T'(x) = \int_0^1 K(s,t)\, f_x'(t,x(t))\cdot dt$ is completely continuous for any $x(s) \in C(0,1)$. $T(x)$ also satisfies H-2, i.e., $T(-x) = -T(x)$ if we assume that $f(s,-x) = -f(s,x)$.

Let us make the following additional assumptions:

H-3: $f(s,x)$ is four times differentiable in x, with $|f_x^{iv}|$ bounded, uniformly over $0 \le s \le 1$. $\lim_{x\to 0} \frac{f(s,x)}{x} = f_x'(s,0)$ uniformly on $0 \le s \le 1$. It can be shown that $\frac{f(s,x)}{x}$ is continuous in x, uniformly on $0 \le s \le 1$.

We assume one or the other of the following pair of hypotheses:

H-4a: Sublinearity; i.e. $f_x'(s,x) > 0$, $0 \le s \le 1$, and $xf_x''(s,x) < 0$, $0 \le s \le 1$, $-\infty < x < +\infty$.

H-4b: Superlinearity; i.e. $f_x'(s,x) > 0$, $0 \le s \le 1$, and $xf_x''(s,x) > 0$, $0 \le s \le 1$, $-\infty < x < +\infty$.

We again note that $f_x'''(s,0) < 0$ in H-4a; $f_x'''(s,0) > 0$ in H-4b; $0 \le s \le 1$. Also for most of our considerations it is well to assume the following:

H-5: Asymptotic Linearity: $\lim_{|x|\to\infty} \frac{f(s,x)}{x} = A(s) \ge 0$ uniformly, $0 \le s \le 1$.

Finally all subsequent considerations are based on the following requirement:

<u>H-6</u>: $K(s,t)$ is a symmetric oscillation kernel, (see section 5).

Hypothesis H-6 together with the condition $f_x'(s,x) > 0$ stated in H-4 imply that the linearized problem

$$\lambda h(s) = \int_0^1 K(s,t) f_x'(t,y(t)) h(t) dt \tag{6.1}$$

possesses a sequence $\{\mu_n\}$ of eigenvalues, each of unit multiplicity and unit Riesz index, and a corresponding sequence $\{h_n\}$ of continuous eigenfunctions such that $h_n(s)$ has precisely n nodes (odd order zeros) and no other zeros on the open interval $0 < s < 1$, $n = 0, 1, 2, \cdots$, [ref. 9, p. 262]. We stress that this property holds whatever the function $y(t) \in C(0,1)$ we choose to substitute in the definition of the operator in eq. (6.1).

Having made all these assumptions, it is clear that every eigenvalue $\mu_p^{(0)}$, $p = 0, 1, 2, \cdots$ is a primary bifurcation point for the nonlinear eigenvalue problem

$$\lambda x(s) = \int_0^1 K(s,t) f(t,x(t)) dt. \tag{6.2}$$

Here $\{\mu_n^{(0)}\}$ is the sequence of simple eigenvalues for linearized eq. (6.1) with $y(t) \equiv 0$. Indeed Theorem 4.2 is applicable, and there exist exactly two solutions of small norm which branch away from the trivial solution $x(s) \equiv \theta$ of eq. (6.2) at $\lambda = \mu_p^{(0)}$, $p = 0, 1, 2, \cdots$. Also, as discussed in section 5, in the sublinear case (H-4a) these solutions branch to the left since $\mu_p^{(o)} > 0$; i.e. there exist two small solutions if $\lambda < \mu_p^{(o)}$, but none if $\lambda > \mu_p^{(o)}$. In the superlinear case on the other hand (H-4b), the branching is to the right, i.e., there exist two small solutions if $\lambda > \mu_p^{(o)}$, but none if $\lambda < \mu_p^{(o)}$. The two solutions of small norm which bifurcate at $\lambda = \mu_p^{(o)}$, $p = 0, 1, 2, \cdots$, differ only in sign. We denote the two solutions bifurcating at $\lambda = \mu_p^{(o)}$ by $x_p^{\pm}(s,\lambda)$, and note that $\lim_{\lambda \to \mu_p^{(o)}} \|x_p^{\pm}(s,\lambda)\| = 0$ in the norm of $C(0,1)$. This is readily seen in inspecting the proof of Theorem 4.2.

The following result on the zeros of $x_p^{\pm}(s,\lambda)$ will be useful:

Theorem 6.1: $x_p^{\pm}(s,\lambda)$, where defined for $\lambda \neq \mu_p^{(o)}$, has exactly p nodes and no other zeros on $0 < s < 1$, $p = 0, 1, 2, \cdots$.

Proof: Consider the problem

$$\mu u(s) = \int_0^1 K(s,t) \frac{f(t,x_p^{\pm}(t,\lambda))}{x_p^{\pm}(t,\lambda)} u(t)dt, \tag{6.3}$$

which has the eigenvalue sequence $\{\mu_n\}$ and eigenfunction sequence $\{u_n(s)\}$ where, as indicated for oscillation kernels, $u_p(s)$ has exactly p nodes on $0 < s < 1$. To convert eq. (6.3) to a problem with a symmetric kernel with the same eigenvalues, we put $V(s) = \sqrt{\frac{f(s,x_p^{\pm}(s,\lambda))}{x_p^{\pm}(s,\lambda)}}\, u(s)$, whence

$$\mu V(s) = \int_0^1 \sqrt{\frac{f(s,x_p^{\pm}(s,\lambda))}{x_p^{\pm}(s,\lambda)}}\, K(s,t) \sqrt{\frac{f(s,x_p^{\pm}(s,\lambda))}{x_p^{\pm}(s,\lambda)}}\, V(t)dt. \tag{6.4}$$

(We note that $\frac{f(s,x)}{x} > 0$.) By H-3 as $\lambda \to \mu_p^{(o)}$, $\lambda \neq \mu_p^{(o)}$, the symmetric kernel tends uniformly to the symmetric kernel $\sqrt{f_x'(s,o)}\,K(s,t)\,\sqrt{f_x'(t,o)}$. Therefore by a known result [ref. 7, p. 151], the eigenvalue μ_p of eq. (6.3) tends to $\mu_p^{(o)}$, $p = 0, 1, 2, \cdots$, and the normalized eigenfunction $V_p(s)$ of eq. (6.4) tends uniformly to $W_p(s)$, where $W_p(s)$ is the p'th normalized eigenfunction of the problem

$$\mu W(s) = \int_0^1 \sqrt{f_x'(s,o)}\,K(s,t)\,\sqrt{f_x'(t,o)}\,W(t)dt$$

which is associated with eigenvalue $\mu_p^{(o)}$. Equivalently we may write

$$\mu_p^{(o)} \frac{W_p(s)}{\sqrt{f_x'(s,o)}} = \int_0^1 K(s,t)f_x'(t,o) \frac{W_p(t)}{\sqrt{f_x'(t,o)}} dt,$$

$$p = 0,1,2,\cdots.$$

But obviously we then have $\frac{W_p(s)}{\sqrt{f_x'(s,o)}} = h_p^{(o)}(s)$, where $h_p^{(o)}(s)$ is the p'th eigenfunction of eq. (6.1) with $y(s) \equiv 0$. This is because the kernel $K(s,t)f_x'(t,o)$ has eigenvalues of unit multiplicity only.

We happen to know a solution pair $(u(s),\mu)$ for eq. (6.3) however, namely $u(s) = \frac{x_p^{\pm}(s,\lambda)}{\|x_p^{\pm}(s,\lambda)\|}$, $\mu = \lambda$. We readily see this by inspection of eq. (6.2). Indeed λ is <u>one</u> of the eigenvalues $\{\mu_n\}$ and $\frac{x_p^{\pm}(x,\lambda)}{\|x_p^{\pm}(s,\lambda)\|}$ is <u>among</u> the normalized eigenfunctions $\{u_n\}$ of eq. (6.3). As $\lambda \to \mu_p^{(o)}$ however, only <u>one</u> of the eigenvalues of eq. (6.3) tends to $\mu_p^{(o)}$, and this must be λ itself. Hence $\lambda = \mu_p$. The corresponding eigenfunction $u_p(s)$

is then a member of the one-dimensional eigenspace spanned by $\dfrac{x_p^{\pm}(s,\lambda)}{\|x_p^{\pm}(s,\lambda)\|}$.

Since $u_p(s)$ has p nodes and no other zeros on $0 < s < 1$, the same is true for $x_p^{\pm}(s,\lambda)$. This concludes the proof.

We prove the following result for the sublinear case. The superlinear case is shown in the same way.

<u>Theorem 6.2</u>: Suppose hypothesis H-4a holds, i.e., we have the sublinear case. Let (x_p^*,λ^*) be a solution pair for eq. (6.2) on the p'th branch $x_p^{\pm}(s,\lambda)$. Then $\mu_p^* < \lambda^*$, where μ_p^* is the p'th eigenvalue of eq. (6.1) where we have put $y(s) = x_p^*$.

<u>Proof</u>: Let K be the positive definite operator on $L_2(0,1)$ generated by $K(s,t)$. ($K(s,t)$ is symmetric, and has positive eigenvalues since it is an oscillation kernel.) There exists a unique positive definite square root H, where $K = H \cdot H$. Let us consider the following eigenvalue problem for a symmetric operator:

$$\lambda \ell = H f'_{x_p^*} H \ell \tag{6.5}$$

with eigenvalue parameter λ and eigenfunctions ℓ to be determined. Eq. (6.5) has the <u>same</u> eigenvalues $\{\mu_n^*\}$ as equation (6.1) with $y(s) = x_p^*$. (To elucidate our notation in eq. (6.5), we state that if the operator H has a kernel $H(s,t)$, then

$$H f'_{x_p^*} H \ell = \int_0^1 H(s,r) f_x'(r, x_p^*(r,\lambda^*)) \int_0^1 H(r,t)\ell(t)\,dt\,dr;$$

the operator $H \dfrac{f}{x_p^*} H$ below would have a corresponding expression.)

Likewise the problem

$$\delta z = H \frac{f}{x_p^*} Hz, \tag{6.6}$$

with a symmetric operator, has the <u>same</u> eigenvalues $\{\delta_n\}$ as problem (6.3) with $x_p^{\pm}(s,\lambda) = x_p^*$; the eigenfunctions of equation (6.6) are used later in the proof, and are denoted by $\{z_n\}$.

We note now that for all $u \in L_2(0,1)$ we have

$$(Hf'_{x_p^*} Hu,u) = (f'_{x_p^*} Hu,Hu) = \int_0^1 \{Hu\}^2 f_x'(s,x_p^*(s,\lambda^*))ds <$$
$$< \int_0^1 \{Hu\}^2 \frac{f(s,x_p^*(s,\lambda^*))}{x_p^*(s,\lambda^*)} ds = (H \frac{f}{x_p^*} Hu,u); \tag{6.7}$$

(in the sublinear case, H-4a, $f(s,x)$ has the property $f_x'(s,x) < \frac{f(s,x)}{x}$, $0 < x < +\infty$ and $-\infty < x < 0$). Then, using Courant's minimax principle [ref. 19, p. 238],

$$\mu_p^* = \min_{\{v_1,\cdots v_{p-1}\}} \max_{u \perp v_1,\cdots,v_{p-1}} \frac{(Hf'_{x_p^*} Hu,u)}{(u,u)}$$
$$\leq \max_{u \perp z_1,\cdots,z_{p-1}} \frac{(Hf'_{x_p^*} Hu,u)}{(u,u)} < \max_{u \perp z_1,\cdots,z_{p-1}} \frac{(H \frac{f}{x_p^*} Hu,u)}{(u,u)} = \delta_p,$$

where δ_p is of course the p'th eigenvalue of eq. (6.3) with $x_p^{\pm}(s,\lambda) = x_p^*$, and we have used inequality (6.7).

Since $K(s,t)$ is an oscillation kernel, δ_p corresponds to that eigenfunction $u_p(s)$ of eq. (6.3) (with $x_p^{\pm}(s,\lambda) = x_p^*$) which has exactly p nodes on $0 < s < 1$. Therefore u_p, $\|u_p\| = 1$ is identical with $\frac{x_p^*(s,\lambda^*)}{\|x_p^*(s,\lambda^*)\|}$, and $\delta_p = \lambda^*$. Hence $\mu_p^* < \lambda^*$ and the theorem is proven.

In the superlinear case, it can be seen from H-4b that $f_x'(s,x) > \frac{f(s,x)}{x}$, $0 < x < +\infty$ and $-\infty < x < 0$; thus for the superlinear case, the inequality of Theorem 6.2 is reversed: $\mu_p^* > \lambda^*$.

As discussed in Section 5, the problem of extending the branches $x_p(s,\lambda)$ into the large is that of finding an ordinary point (i.e. (x_p,λ) such that $\lambda I - Kf'_{x_p}$ has a bounded inverse, where Kf'_{x_p} is the operator of eq. (6.1)), employing the process of Theorem 2.3 in a stepwise manner, which process can terminate only in an exceptional point by Theorem 2.4, and finally handling the bifurcation equation (3.8) at any exceptional points on the branch. Theorem 6.2 above represents about half of the task of showing that $x_p^* = x_p^{\pm}(s,\lambda^*)$ is an ordinary point. We have shown that $\mu_p^* < \lambda^*$ ($\mu_p^* > \lambda^*$) there, and that this is a consequence of sublinearity(superlinearity) and thus is true wherever the branch is developed. Note that because $K(s,t)$ is an oscillation kernel, μ_p^* is of unit multiplicity and so $\mu_n^* < \mu_p^*$, $n = p+1, p+2, \cdots$ for arbitrary points x_p^* on the p'th branch $x_p^{\pm}(s,\lambda)$.

The other half of the job of showing that $x_p^* = x_p^{\pm}(s,\lambda^*)$ is an ordinary point consists in showing that $\lambda^* < \mu_{p-1}^*$, where μ_{p-1}^* is of course the p-1th eigenvalue of eq. (6.1) with $y(t) = x_p^*$. Indeed, if we can show that $\lambda^* < \mu_{p-1}^*$, then $\cdots \mu_{p+1}^* < \mu_p^* < \lambda^* < \mu_{p-1}^* < \mu_{p-2}^* \cdots$ so that $\lambda^* I - Kf'_{x_p^*}$ cannot be singular.

Showing that $\lambda^* < \mu_{p-1}^*$ presents major difficulties; further assumptions are needed. So as to introduce the necessary assumptions, we prove the following intermediate results:

<u>Lemma 6.3</u>: Consider the linear integral operator

$$K\varphi h = \int_0^1 K(s,t)\varphi(t)h(t)dt \tag{6.8}$$

where $\varphi \in C(0,1)$, $\varphi \geq 0$, and $K(s,t)$ is an oscillation kernel. The eigenvalues $\{\mu_n(K,\varphi)\}$ are continuous functions of φ in terms of the $L_2(0,1)$ norm.

Proof: By putting $\psi = \sqrt{\varphi}\, h$, we ascertain that the problem $\lambda\psi = \sqrt{\varphi}\, K \sqrt{\varphi}\, \psi$ has the same eigenvalue parameter λ as the problem $\lambda h = K\varphi h$ for the operator defined in eq. (6.8). Here of course we have the symmetric operator

$$\sqrt{\varphi}\, K \sqrt{\varphi}\, \psi = \int_0^1 \sqrt{\varphi(s)}\, K(s,t) \sqrt{\varphi(t)}\, \psi(t)dt$$

with $\varphi(s) \geq 0$. As $\varphi \to \varphi^*$, $\varphi, \varphi^* \in C(0,1)$, in the $L_2(0,1)$ norm, we have

$$\|\sqrt{\varphi}\, K \sqrt{\varphi} - \sqrt{\varphi^*}\, K \sqrt{\varphi^*}\|$$

$$\leq \|\sqrt{\varphi}\, K \sqrt{\varphi} - \sqrt{\varphi^*}\, K \sqrt{\varphi}\| + \|\sqrt{\varphi^*}\, K \sqrt{\varphi} - \sqrt{\varphi^*}\, K \sqrt{\varphi^*}\|$$

$$\leq \left[\int_0^1 \int_0^1 |\sqrt{\varphi(s)} - \sqrt{\varphi^*(s)}|^2 K^2(s,t)|\varphi(t)|\,dtds\right]^{\frac{1}{2}}$$

$$+ \left[\int_0^1 \int_0^1 |\varphi^*(s)|K^2(s,t)|\sqrt{\varphi(t)} - \sqrt{\varphi^*(t)}|^2 dtds\right]^{\frac{1}{2}}$$

$$\leq \left[\int_0^1 \int_0^1 |\varphi(s)-\varphi^*(s)|K^2(s,t)|\varphi(t)|\,dtds\right]^{\frac{1}{2}}$$

$$+ \left[\int_0^1 \int_0^1 |\varphi^*(s)|K^2(s,t)|\varphi(t)-\varphi^*(t)|\,dtds\right]^{\frac{1}{2}}$$

$$\leq \bar{K}\|\varphi-\varphi^*\| \cdot (\|\varphi\|+\|\varphi^*\|) \to 0,$$

where $\bar{K} = \max\limits_{\substack{0\leq s\leq 1\\ 0\leq t\leq 1}} K(s,t)$. Thus $\sqrt{\varphi}\, K \sqrt{\varphi} \to \sqrt{\varphi^*}\, K \sqrt{\varphi^*}$ in the uniform topology of $L_2(0,1)$. Since $\sqrt{\varphi}\,K\,\sqrt{\varphi}$ is a symmetric operator, we use a known result [ref. 19, p. 239] to see that $\mu_n(K,\varphi) \to \mu_n(K,\varphi^*)$ as $\varphi \to \varphi^*$ in the $L_2(0,1)$ norm, $n = 0,1,2,\cdots$. Here of course $\mu_n(K,\varphi)$ is the n'th eigenvalue of operator (6.8). This ends the proof.

Next we define the following number which varies from branch to branch:

$$\Lambda_p = \sup_{\varphi \in S_1^+} \frac{\mu_p(K,\varphi)}{\mu_{p-1}(K,\varphi)}$$

where $S_1^+ = [\varphi | \varphi \in C(0,1),\ \|\varphi\| = 1,\ \varphi \geq 0]$. Of course $\mu_p(K,\varphi)$ is the p'th eigenvalue of operator (6.8). We have the following result:

<u>Lemma 6.4</u>: $\Lambda_p > 0$ is less than unity, and the maximum is actually assumed for a function $\varphi^* \in S_1^+$.

<u>Proof</u>: The positive symmetric continuous nondegenerate kernel $K(s,t)$ generates a nonsingular completely continuous operator K on $L_2(0,1)$. Thus $\lambda = 0$ is in the continuous spectrum of K on $L_2(0,1)$, and the range R_K of K is dense in $L_2(0,1)$, [ref. 21, p. 305, Th. 8; p. 292]. We have $R_K \subset C\,(0,1)$ since $K(s,t)$ is continuous. Also of course $S_1^+ \subset C(0,1) \subset L_2(0,1)$ in the sense of set-inclusion.

Now $\Lambda_p = \sup\limits_{\varphi \in S_1^+ \cap R_K} \dfrac{\mu_p(K,\varphi)}{\mu_{p-1}(K,\varphi)}$ since $F_p(\varphi) = \dfrac{\mu_p(K,\varphi)}{\mu_{p-1}(K,\varphi)}$ is strongly continuous in $\varphi \in S_1^+$ in terms of the $L_2(0,1)$ norm, as was shown in Lemma 6.3. Then

$$\Lambda_p = \sup_{\varphi \in S_1^+ \cap R_K} F_p(\varphi) \leq \sup_{\substack{g \in L_2(0,1) \\ Kg \geq 0}} F_p(Kg) = \sup_{g \in E_K} F_p(Kg) = \Lambda_p' \leq 1,$$

where $E_K = [g | g \in L_2(0,1), \|g\| = 1, Kg \geq 0]$. This follows since the functional $F_p(Kg)$ is constant along rays rg, $0 \leq r < \infty$.

Instead of continuing to deal directly with the functional $F_p(Kg)$, it is advantageous at this point to consider the functional $\Phi_p^R(g) = 1+F_p(Kg)$ on E_K, and its convenient extension $\Phi_p(g) = (g,g) + F_p(Kg)$ to all g such that $Kg \geq 0$, $g \in L_2(0,1)$. If Φ_p^R assumes its maximum value $1+\Lambda_p'$ on E_K, then Λ_p' will be assumed by $F_p(Kg)$ at the same element of E_K.

There exists a maximizing sequence $\{g_n\} \in E_K$ such that $\Phi_p^R(g_n) > 1 + \Lambda_p' - \frac{1}{n}$, $n = 1,2,3,\cdots$, and $\lim_{n\to\infty} \Phi_p^R(g_n) = \sup_{g\in E_K} \Phi_p^R(g) = 1 + \Lambda_p'$. Since the unit sphere is weakly compact in $L_2(0,1)$, there exists a subsequence $\{g_{n_1}\}$, $\|g_{n_1}\| = 1$, weakly convergent to some $g^* \in L_2(0,1)$. By passing to the weak limit in the inequality $|(g_{n_1}, g^*)| \leq \|g_{n_1}\|\|g^*\|$ as $n_1 \to \infty$, we see that $\|g^*\| \leq 1$.

Since K is compact, $\{Kg_{n_1}\}$ converges strongly in $L_2(0,1)$ to an element $\varphi^* = Kg^* \geq 0$. Because $F_p(\varphi)$ is a continuous function of $\varphi \in C(0,1)$ in the $L_2(0,1)$ norm by Lemma 6.3, we have $F_p(Kg_{n_1}) \to \Lambda_p'$ as $n_1 \to \infty$. In the case $g^* \not\equiv 0$ therefore, the maximum $1 + \Lambda_p'$ of $\Phi_p^R(g)$ on E_K is assumed by $\Phi_p(g)$ at g^*. If $\|g^*\| < 1$, we should then have $\Phi_p(g^*) = (g^*,g^*) + F_p(Kg^*)$ $= 1 + \Lambda_p'$. This is a contradiction since $\Phi_p(g^*) < \Phi_p^R\left(\frac{g^*}{\|g^*\|}\right)$, which in turn is because $F_p(Kg)$ is constant on radial lines, and $\frac{g^*}{\|g^*\|} \in E_K$. Hence $\|g^*\| = 1$, and both $\Phi_p^R(g)$ and $F_p(Kg)$ assume their maximum values, $1 + \Lambda_p'$ and Λ_p' respectively, on E_K at the element $g^* \in E_K$. The maximum value of $F_p(\varphi)$ on S_1^+ is therefore assumed at $\frac{\varphi^*}{\|\varphi^*\|}$, where $\varphi^* = Kg^* \geq 0$, $\varphi^* \not\equiv 0$.

Also since K(s,t) is a continuous kernel, we have $\varphi^*(s) \in C(0,1)$, so that $\Lambda_p{}' = \Lambda_p$.

In the case $g^* = 0$, $F_p(Kg^*)$ is not defined, but the limiting value $\Lambda_p{}'$ of $\Phi_p(g)$ as $g_{n_1} \to g^*$ weakly in $L_2(0,1)$ is less than certain discernable values assumed by $\Phi_p(g)$ on E_K, which is a contradiction.

The linear operator (6.8), with $\varphi \equiv \varphi^*$, has eigenvalues $\mu_n(K,\varphi^*)$, $n = 0,1,2,\cdots$, such that $\mu_n(K,\varphi^*) < \mu_{n-1}(K,\varphi^*)$ (strict inequality); this is because K(s,t) is an oscillation kernel, [ref. 9, pp. 254-273]. Hence $\Lambda_p < 1$, and the lemma is proven.

With these two preliminary results proven, we are now in a position to state our main results having to do with whether or not a given point $x_p^* = x_p^{\pm}(s,\lambda^*)$ on the p'th branch is an ordinary point.

We put forth a couple of conditions which, together with the statement of Theorem 6.2, will be seen to guarantee that x_p^* is an ordinary point. These may be considered as a priori conditions on either the kernel K(s,t) or on the function f(s,x). In the sublinear case, hypothesis H-4a, the condition is that

$$\frac{xf_x'(s,x)}{f(s,x)} > \Lambda_p, \quad 0 \le s \le 1, \ -\infty < x < \infty, \tag{6.9a}$$

while in the superlinear case, hypothesis H-4b the condition is that

$$\frac{xf_x'(s,x)}{f(s,x)} < \frac{1}{\Lambda_{p+1}}, \quad 0 \le s \le 1, \ -\infty < x < +\infty, \tag{6.9b}$$

where Λ_n, $n = 1,2,3,\cdots$ was shown in Lemma 6.4 to be less than unity. The fact that $0 < \Lambda_n < 1$, $n = 1,2,3,\cdots$ means that there exists a class of sublinear respectively superlinear functions f(s,x) for which condition (6.9a) or (6.9b) is realizable.

Theorem 6.5: Under hypothesis H-4a (sublinearity) let $f(s,x)$ be such that condition (6.9a) is satisfied. Then $\lambda^* < \mu^*_{p-1}$, where μ^*_{p-1} is the p-1th eigenvalue of eq. (6.1) where we have put $y(s) = x_p^* = x_p^{\pm}(s,\lambda^*)$. Under hypothesis H-4b (superlinearity) let $f(s,x)$ be such that condition (6.9b) is satisfied. Then $\mu^*_{p+1} < \lambda^*$.

Proof: As in the proof of Theorem 6.2 we again split the kernel: $K = H \cdot H$, and consider the symmetrized problem

$$\delta \ell = H \frac{f}{x_p^*} H\ell, \tag{6.10}$$

where, if the square root operator H has a kernel $H(s,t)$, we may write

$$H \frac{f}{x_p^*} H\ell = \int_0^1 H(s,r) \frac{f(r,x_p^*(r,\lambda^*))}{x_p^*(r,\lambda^*)} \int_0^1 H(r,t)\ell(t)dtdr.$$

For the p'th eigenvalue of eq. (6.10) we have $\delta_p = \lambda^*$. Now in the sublinear case (Hypothesis H-4a),

$$\lambda^* = \delta_p = \mu_p\left(K, \frac{f}{x_p^*}\right) = \frac{\mu_p\left(K, \frac{f}{x_p^*}\right)}{\mu_{p-1}\left(K, \frac{f}{x_p^*}\right)} \mu_{p-1}\left(K, \frac{f}{x_p^*}\right) \le \Lambda_p \mu_{p-1}\left(K, \frac{f}{x_p^*}\right) \tag{6.11}$$

as follows from the definition of Λ_p. But

$$\mu_{p-1}\left(K, \frac{f}{x_p^*}\right) = \min_{\{v_1,\cdots,v_{p-2}\}} \max_{u \perp v_1,\cdots,v_{p-2}} \frac{\left(H \frac{f}{x_p^*} Hu, u\right)}{(u,u)}$$

$$\le \max_{u \perp y_1,\cdots,y_{p-2}} \frac{\left(H \frac{f}{x_p^*} Hu, u\right)}{(u,u)}$$

where $y_1, \cdots, y_{p-2}$ can be any set of p-2 elements in $L_2(0,1)$. In particular, let $y_1, \cdots, y_{p-2}$ be the eigenelements of $Hf'_{x_p^*} H$, which operator was discussed in the proof of Theorem 6.2. Then since by condition (6.9a),

$$\Lambda_p \left(H \frac{f}{x_p^*} Hu, u \right) = \Lambda_p \left(\frac{f}{x_p^*} Hu, Hu \right)$$

$$= \Lambda_p \int_0^1 \frac{f(s, x_p^*(s,\lambda^*))}{x_p^*(s,\lambda^*)} \{Hu\}^2 ds < \int_0^1 f_x'(s, x_p^*(s,\lambda^*)) \{Hu\}^2 ds \tag{6.12}$$

$$= (Hf'_{x_p^*} Hu, u)$$

we can write, using inequality (6.11)

$$\lambda^* \leq \Lambda_p \mu_{p-1} \left(K, \frac{f}{x_p^*} \right) \leq \max_{u \perp y_1, \cdots, y_{p-2}} \Lambda_p \frac{\left(H \frac{f}{x_p^*} Hu, u \right)}{(u,u)}$$

$$< \max_{u \perp y_1, \cdots, y_{p-2}} \frac{(Hf'_{x_p^*} Hu, u)}{(u,u)} = \mu^*_{p-1} ,$$

which proves the result in the sublinear case.

On the other hand, in the superlinear case (Hypothesis H-4b),

$$\mu_{p+1} (K, f'_{x_p^*}) = \frac{\mu_{p+1} (K, f'_{x_p^*})}{\mu_p (K, f'_{x_p^*})} \mu_p (K, f'_{x_p^*}) < \Lambda_{p+1} \mu_p (K, f'_{x_p^*}) \tag{6.13}$$

where again we use the definition of the number Λ_n. Analogously with inequality (6.12), we now have the inequality

$$\Lambda_{p+1}(Hf'_{x_p^*} Hu,u) = \Lambda_{p+1}(f'_{x_p^*} Hu,Hu)$$

$$= \Lambda_{p+1} \int_0^1 f_x'(s,x_p^*(s,\lambda^*))\{Hu\}^2 ds$$

$$< \int_0^1 \frac{f(s,x_p^*(s,\lambda^*))}{x_p^*(s,\lambda^*)} \{Hu\}^2 ds = \left(H \frac{f}{x_p^*} Hu,u\right) \quad (6.14)$$

where we have made use of condition (6.9b). Then using inequality (6.13) we have, with $z_1,\cdots,z_{p-1}$ representing eigenelements of $H \frac{f}{x_p^*} H$,

$$\mu_{p+1} < \Lambda_{p+1} \min_{\{v_1,\cdots,v_{p-1}\}} \max_{u\perp v_1,\cdots,v_{p-1}} \frac{(Hf'_{x_p^*} Hu,u)}{(u,u)}$$

$$< \Lambda_{p+1} \max_{u\perp z_1,\cdots,z_{p-1}} \frac{(Hf'_{x_p^*} Hu,u)}{(u,u)} < \max_{u\perp z_1,\cdots,z_{p-1}} \frac{\left(H \frac{f}{x_p^*} Hu,u\right)}{(u,u)}$$

$$= \mu_p\left(K, \frac{f}{x_p^*}\right) = \delta_p = \lambda^*,$$

where inequality (6.14) has been used. This proves the theorem.

The basic content of Theorem 6.2 and Theorem 6.5 taken together can be stated quite vividly:

<u>Corollary 6.6</u>: In the sublinear case (Hypothesis H-4a), if $\Lambda_p < \frac{xf_x'(s,x)}{f(s,x)} < 1$, $0 \le s \le 1$, $-\infty < x < +\infty$, then $\mu_p < \lambda < \mu_{p-1}$. Here μ_p is the p'th eigenvalue of the operator $Kf_x'h = \int_0^1 K(s,t)f_x'(t,x_p^{\pm}(t,\lambda))h(t)dt$, where $x_p^{\pm}(t,\lambda)$ represents the p'th branch of the eigenfunctions of the nonlinear problem. On the other hand, in the superlinear case (Hypothesis H-4b), if $1 < \frac{xf_x'(s,x)}{f(s,x)} < \frac{1}{\Lambda_{p+1}}$, $0 \le s \le 1$, $-\infty < x < +\infty$, then $\mu_{p+1} < \lambda < \mu_p$.

Thus, since $(\lambda I-Kf_x')^{-1}$ can fail to exist as a bounded inverse only when $\lambda = \mu_n$ for some $n = 0,1,2,\cdots$, and since under the assumption that we have an oscillation kernel (Hypothesis H-6) all the μ_n's are a priori of unit multiplicity, we have stated in Corollary 6.6 certain a priori conditions that we never must consider the bifurcation equation, eq.(4.1), in developing the p'th branch of eigenfunctions $x_p^{\pm}(s,\lambda)$ of the Hammerstein operator from the small into the large. In essence we have opened up a corridor between the eigenvalues $\{\mu_n\}$ through which to extend the p'th branch uniquely.

The definition of the number Λ_p can certainly be refined. This quantity was used in the proof of Theorem 6.5 (inequalities 6.11 and 6.13) to represent an upper bound for the ratios $\dfrac{\mu_p\left(K,\frac{f}{x_p^*}\right)}{\mu_{p-1}\left(K,\frac{f}{x_p^*}\right)}$ and $\dfrac{\mu_p(K,f'_{x_p^*})}{\mu_{p-1}(K,f'_{x_p^*})}$. In the definition of Λ_p (Lemma 6.4) we took the sup of the ratio $\dfrac{\mu_p(K,\varphi)}{\mu_{p-1}(K,\varphi)}$ as φ assumed values in the set $S_1^+ = [\varphi|\varphi\in C(0,1),\ \|\varphi\| = 1,\ \varphi \ge 0]$. Clearly however it would have been better to have taken the sup over a more restricted set of functions which would have reflected somehow the property that functions on the p'th branch $x_p^* = x_p^{\pm}(s,\lambda^*)$ have exactly p nodes on the open interval $0 < s < 1$, (see Theorem 6.1).

To begin with, we see that in the sublinear case (H-4a) we have the inequalities $A(s) \le \frac{f}{x_p^*} \le f_x'(s,0)$ and $A(s) \le f'_{x_p^*} \le f_x'(s,0)$, while in the superlinear case $f_x'(s,0) \le \frac{f}{x_p^*} \le A(s)$ and $f_x'(s,0) \le f'_{x_p^*} \le A(s)$.

Here A(s) is the function appearing in the statement of Hypothesis H-5. Hence the set over which we compute the sup in the definition of Λ_p should reflect this fact.

We confine ourselves to the sublinear case (H-4a), the superlinear case being handled in a similar way. Let

$$S^{+}_{A,f_0'} = [\varphi|\varphi \in C(0,1),\ A(s) \le \varphi(s) \le f_x'(s,0)],$$

and consider the cone in S^+ all of whose rays intersect this convex set:

$$C_{A,f_0'} = [\psi|\lambda\psi(s) \in S^{+}_{A,f_0'} \quad \text{for some } \lambda,\ \lambda > 0].$$

Then we may use

$$\check{\Lambda}_p = \sup_{\varphi \in S_1^+ \cap C_{A,f_0'}} \frac{\mu_p(K,\varphi)}{\mu_{p-1}(K,\varphi)} \le \Lambda_p < 1$$

as a refined definition of the number Λ_p to be used in inequalities (6.11) and (6.13) (the latter in the superlinear case).

We can refine further the definition of Λ_p. Let

$$S^{+}_{A,f_0',p} = [\varphi|\varphi \in C(0,1),\ A(s) \le \varphi(s) \le f_x'(s,0);\ f_x'(s,0) - \varphi(s) \text{ has exactly } p \text{ zeros on } 0 < s < 1]$$

and then consider the cone in S^+ all of whose rays intersect this set:

$$C_{A,f_0',p} = [\psi|\lambda\psi(s) \in S^{+}_{A,f_0',p} \quad \text{for some } \lambda,\ \lambda > 0].$$

Then we may employ

$$\check{\check{\Lambda}}_p = \sup_{\varphi \in S_1^+ \cap C_{A,f_0',p}} \frac{\mu_p(K,\varphi)}{\mu_{p-1}(K,\varphi)} \le \check{\Lambda}_p \le \Lambda_p < 1$$

as a yet more refined definition which takes into account that in inequalities (6.11) and (6.13) we are really only interested in the p'th branch. It is not known at present to what extent, if any, such refined definitions of Λ_p improve the sufficient conditions against secondary bifurcation, of the p'th branch of eigenfunctions, given in Theorem 6.5 and in Corollary 6.6. It is clear that such refined definitions do not enhance the computability of Λ_p.

Finally an examination of the proof of Theorem 6.5 will suggest another condition against secondary bifurcation of the p'th branch of eigenfunctions of eq. (6.2), which is much less concrete than that given by inequalities (6.9). Namely, it is a condition which compares two functionals on $S^+ = [\varphi | \varphi \in C(0,1), \varphi \geq 0]$. In the sublinear case, hypothesis H-4a, the condition is that

$$\min_{0\leq s\leq 1} \frac{\varphi f_x'(s,\varphi)}{f(s,\varphi)} > \frac{\mu_p\left(K, \frac{f(s,\varphi)}{\varphi}\right)}{\mu_{p-1}\left(K, \frac{f(s,\varphi)}{\varphi}\right)} \qquad \varphi \in S^+ \tag{6.15a}$$

while in the superlinear case, hypothesis H-4b, the condition is that

$$\max_{0\leq s\leq 1} \frac{\varphi f_x'(s,\varphi)}{f(s,\varphi)} < \frac{\mu_p(K, f_x'(s,\varphi))}{\mu_{p+1}(K, f_x'(s,\varphi))} \qquad \varphi \in S^+. \tag{6.15b}$$

(We remember that $\frac{f(s,\varphi)}{\varphi}$ and $f_x'(s,\varphi)$ are even functions of φ, so that an oscillatory $\varphi(s)$ would satisfy the inequality in (6.15) along with $|\varphi(s)|$.)

In view of Lemma 6.4, the right hand functional in inequality (6.15a) is less than unity, while the right hand functional in equality (6.15b) exceeds unity. Hence the conditions are realizable for sublinear respectively superlinear functions $f(s,x)$.

Theorem 6.7: Under hypothesis H-4a (sublinearity) let $f(s,x)$ be such that condition (6.15a) is satisfied. Then $\lambda^* < \mu^*_{p-1}$ where μ^*_{p-1} is the p-1th eigenvalue of eq. (6.1) with $y(s) = x_p{}^* = x_p{}^{\pm}(s,\lambda^*)$. Again under hypothesis H-4b (superlinearity), let $f(s,x)$ be such that condition (6.15b) is satisfied). Then $\mu^*_{p+1} < \lambda^*$.

Proof: We show the proposition in the sublinear case, the superlinear case being shown similarly. We split the kernel: $K = H\cdot H$, and have eq. (6.10). Then from eq. (6.11) we have

$$\lambda^* = \frac{\mu_p\left(K, \frac{f}{x_p{}^*}\right)}{\mu_{p-1}\left(K, \frac{f}{x_p{}^*}\right)} \mu_{p-1}\left(K, \frac{f}{x_p{}^*}\right)$$

$$\leq \max_{u \perp y_1, \cdots y_{p-2}} \frac{\left(H \dfrac{\mu_p\left(K, \frac{f}{x_p{}^*}\right)}{\mu_{p-1}\left(K, \frac{f}{x_p{}^*}\right)} \dfrac{f}{x_p{}^*} Hu, u\right)}{(u,u)} \tag{6.16}$$

where the arbitrary set of p-2 elements $y_1, \cdots, y_{p-2}$ in $L_2(0,1)$ can be the eigenelements of $Hf'_{x_p{}^*} H$. In inequality (6.16) we have used the fact that $\dfrac{\mu_p\left(K, \frac{f}{x_p{}^*}\right)}{\mu_{p-1}\left(K, \frac{f}{x_p{}^*}\right)}$ is a pure number. Then in view of condition (6.15a), we can write, as in inequality (6.12):

$$\left(H \frac{\mu_p\left(K, \frac{f}{x_p^*}\right)}{\mu_{p-1}\left(K, \frac{f}{x_p^*}\right)} \frac{f}{x_p^*} Hu, u\right) = \left(\frac{\mu_p\left(K, \frac{f}{x_p^*}\right)}{\mu_{p-1}\left(K, \frac{f}{x_p^*}\right)} \frac{f}{x_p^*} Hu, Hu\right)$$

$$= \int_0^1 \frac{\mu_p\left(K, \frac{f}{x_p^*}\right)}{\mu_{p-1}\left(K, \frac{f}{x_p^*}\right)} \frac{f(s,x_p^*(s,\lambda^*))}{x_p^*(s,\lambda^*)} \{Hu\}^2 ds$$

$$< \int_0^1 \frac{x_p^*(s,\lambda^*) f_x'(s,x_p^*(s,\lambda^*))}{f(s,x_p^*(s,\lambda^*))} \frac{f(s,x_p^*(s,\lambda^*))}{x_p^*(s,\lambda^*)} \{Hu\}^2 ds$$

$$= \int_0^1 f_x'(s,x_p^*(s,\lambda^*)) \{Hu\}^2 ds = (Hf'_{x_p^*} Hu, u).$$

Then using inequality (6.16) we have

$$\lambda^* < \max_{u \perp y_1, \cdots, y_{p-2}} \frac{(Hf'_{x_p^*} Hu, u)}{(u,u)} = \mu^*_{p-1} ,$$

which proves the sublinear case. As mentioned, the superlinear case, utilizing condition (6.15b), is similar and the proof can be patterned after that of Theorem 6.5.

There follows immediately the result:

<u>Corollary 6.8</u>: In the sublinear case (Hypothesis H-4a), if

$$\frac{\mu_p\left(K, \frac{f(s,\varphi)}{\varphi}\right)}{\mu_{p-1}\left(K, \frac{f(s,\varphi)}{\varphi}\right)} < \min_{0 \le s \le 1} \frac{\varphi f_x'(s,\varphi)}{f(s,\varphi)} < 1, \quad \varphi \in S^+ ,$$

then $\mu_p < \lambda < \mu_{p-1}$, and there is no secondary bifurcation of the p'th branch of eigenfunctions of eq. (6.2). In the superlinear case (Hypothesis H-4b), if

$$1 < \max_{0\le s\le 1} \frac{\varphi f_x'(s,\varphi)}{f(s,\varphi)} < \frac{\mu_p(K,f_x'(s,\varphi))}{\mu_{p+1}(K,f_x'(s,\varphi))}, \quad \varphi \in S^+ ,$$

then $\mu_{p+1} < \lambda < \mu_p$, and again there is no secondary bifurcation.

Of course, in this corollary, μ_p is the p'th eigenvalue of the operator $Kf_x'h = \int_0^1 K(s,t)f_x'(t,x_p^{\pm}(t,\lambda))h(t)dt$, where $x_p^{\pm}(t,\lambda)$ represents the p'th branch of eigenfunctions of nonlinear problem (6.2).

It would be interesting to find out to what extent condition (6.15) might be a necessary condition for no secondary bifurcation of the p'th branch. Certainly for Hammerstein operators under hypothesis H-4a (sublinearity), it is necessary for no secondary bifurcation that $\lambda^* < \mu^*_{p-1}$, where again μ^*_{p-1} is the p-1th eigenvalue of eq. (6.1) with $y(s) = x_p^{\pm}(s,\lambda^*)$. Now $\lambda^* < \mu^*_{p-1}$ means that

$$\max_{h\perp z_1,\cdots z_{p-1}} \frac{\left(H \frac{f}{x_p^*} Hh,h\right)}{(h,h)} < \min_{v_1,\cdots v_{p-1}} \max_{h\perp v_1,\cdots,v_{p-1}} \frac{\left(\frac{\mu_{p-1}(K,f'_{x_p^*})}{\mu_p(K,f'_{x_p^*})} Hf'_{x_p^*} Hh,h\right)}{(h,h)}$$

using obvious notation from the proofs of Theorems 6.5 and 6.7. Here $z_1,\cdots,z_{p-1}$ are of course the first p-1 eigenelements of the operator $H \frac{f}{x_p^*} H$. Then for some $\bar{h} \perp z_1, \cdots, z_{p-1}$, we have

$$\max_{h\perp z_1,\cdots z_{p-1}} \frac{\left(H \frac{f}{x_p^*} Hh,h\right)}{(h,h)} < \max_{h\perp z_1,\cdots,z_{p-1}} \frac{\left(\frac{\mu_{p-1}(K,f'_{x_p^*})}{\mu_p(K,f'_{x_p^*})} Hf'_{x_p^*} Hh,h\right)}{(h,h)}$$

$$= \frac{\left(\frac{\mu_{p-1}(K,f'_{x_p^*})}{\mu_p(K,f'_{x_p^*})} Hf'_{x_p^*} H\bar{h},\bar{h}\right)}{(\bar{h},\bar{h})} ,$$

which leads immediately to

$$\frac{\left(H \frac{f}{x_p^*} H\bar{h},\bar{h}\right)}{(\bar{h},\bar{h})} < \frac{\left(\frac{\mu_{p-1}(K,f'_{x_p^*})}{\mu_p(K,f'_{x_p^*})} Hf'_{x_p^*} H\bar{h},\bar{h}\right)}{(\bar{h},\bar{h})} .$$

Thus, cancelling denominators we have

$$\left(H\left\{\frac{f}{x_p^*} - \frac{\mu_{p-1}(K,f'_{x_p^*})}{\mu_p(K,f'_{x_p^*})} f'_{x_p^*}\right\}H\bar{h},\bar{h}\right) < 0$$

or

$$\left(\left\{\frac{f}{x_p^*} - \frac{\mu_{p-1}(K,f'_{x_p^*})}{\mu_p(K,f'_{x_p^*})} f'_{x_p^*}\right\}H\bar{h},H\bar{h}\right) < 0$$

which has the concrete representation

$$\int_0^1 \left\{\frac{f(s,x_p(s,\lambda^*))}{x_p(s,\lambda^*)} - \frac{\mu_{p-1}(K,f'_{x_p^*})}{\mu_p(K,f'_{x_p^*})} f_x'(s,x_p(s,\lambda^*))\right\} \{H\bar{h}\}^2 ds < 0.$$

For this it is necessary that

$$\frac{f(s,x_p(s,\lambda^*))}{x_p(s,\lambda^*)} < \frac{\mu_{p-1}(K,f'_{x_p^*})}{\mu_p(K,f'_{x_p^*})} f_x'(s,x_p(s,\lambda^*))$$

or

$$\frac{x_p(s,\lambda^*)f_x'(s,x_p(s,\lambda^*))}{f(s,x_p(s,\lambda^*))} > \frac{\mu_p(K,f'_{x_p^*})}{\mu_{p-1}(K,f'_{x_p^*})} \qquad (6.17)$$

on a subset of $0 \leq s \leq 1$ of positive measure.

Necessary condition (6.17) certainly bears some resemblance to sufficient condition (6.15a) for the sublinear case, but these two conditions also show a conjugate feature: $\frac{f(s,\varphi)}{\varphi}$ occurs in the sufficient condition while $f'_{x_p^*}$ occurs in the necessary condition.

A necessary condition for the superlinear case is handled in the same way.

7. Extension of Branches of Eigenfunctions of Hammerstein Operators.

As remarked in the text of section 6 prior to introducing Theorem 6.2, Theorem 4.2 is applicable in defining the branches of eigenfunctions of the nonlinear Hammerstein equation, eq. (6.2), in a small neighborhood of the origin and in a neighborhood of a primary bifurcation point. Indeed under Hypotheses H-2 through H-6 there exist exactly two branches $x_p^{\pm}(s,\lambda)$ emanating from the trivial solution at each primary bifurcation point $\lambda = \mu_p^{(o)}$. In the sublinear case (H-4a) these exist for $\lambda < \mu_p^{(o)}$, while in the superlinear case (H-4b) they exist for $\lambda > \mu_p^{(o)}$. By the supposition of oddness (H-2), the two branches $x_p^{\pm}(s,\lambda)$ differ only in sign.

In order to employ considerations of Theorems 2.3 and 2.4 to extend the p'th branch $x_p^{\pm}(s,\lambda)$ from the small into the large, we needed assurance that there existed some ordinary point (x_p^*,λ^*) on that branch, i.e. a point such that $\lambda^*I-T'(x_p^*)$ has a bounded inverse. This assurance is given by Corollary 6.6 <u>or</u> Corollary 6.8 under the assumption that either condition (6.9) <u>or</u> condition (6.15) holds, whether we have sublinearity (H-4a) or superlinearity (H-4b). Moreover, we shall see that either of these corollaries gives assurance that, a priori, all points (x_p,λ) on the p'th branch $x_p^{\pm}(s,\lambda)$ are indeed ordinary points. Of course the latter can be inferred also from Theorem 2.4 once a single ordinary point is found, but there is no assurance that the branch cannot terminate at a singular point on the basis of Theorem 2.4.

Accordingly we invoke Theorem 2.4 and state that there does exist a branch $x_p^{\pm}(s,\lambda)$, or a "unique maximal sheet," of eigenfunctions of problem

(6.2) emanating from the trivial solution at the primary bifurcation point $\lambda = \mu_p^{(o)}$. The only finite boundary point such a sheet may have is a point (x_p^*, λ^*) such that $\lambda^* I - Kf'_{x_p^*}$ has no bounded inverse.

Theorem 7.1: The branch $x_p^{\pm}(s,\lambda)$ has no finite boundary point apart from $x = \theta$, and may therefore be continued indefinitely.

Proof: If there were such a boundary point (x_p^*, λ^*), then $x_p(s,\lambda) \to x_p^*$ as $\lambda \to \lambda^*$ in the $C(0,1)$ norm. By Theorem 6.1, $x_p^*(s,\lambda^*)$ has exactly p nodes on $0 < s < 1$. Accordingly in view of Theorem 6.2 and either Theorem 6.5 or Theorem 6.7 we have $\mu_p^* < \lambda^* < \mu_{p-1}^*$ in the sublinear case and $\mu_{p+1}^* < \lambda^* < \mu_p^*$ in the superlinear case. Hence $\lambda^* I - Kf'_{x_p^*}$ must have a bounded inverse, i.e. λ^* is not an eigenvalue of $Kf'_{x_p^*}$. This however is a contradiction, since by Theorem 2.4, a boundary point is an exceptional point.

Theorem 7.2: There exists a number $\bar{\lambda} \geq 0$ such that $\limsup_{\lambda \to \bar{\lambda}} \|x_p(s,\lambda)\| = \infty$. (Note: In the superlinear case it is possible that $\bar{\lambda} = \infty$.)

Proof: We assume the sublinear case (H-4a), the superlinear case being similar. Let

$$\Pi_p = \{\lambda | 0 < \lambda < \mu_p^{(o)},\ x_p^{\pm}(s,\lambda) \text{ exists and is continuous in } \lambda \text{ uniformly with respect to } 0 \leq s \leq 1\}.$$

By sublinearity (H-4a), Theorem 4.2 and Chapter 5, Π_p includes some small interval $(\mu_p^{(o)} - \epsilon, \mu_p^{(o)})$, a left neighborhood of $\mu_p^{(o)}$. Let us suppose, contrary to the statement of the theorem, that there exists a number $M > 0$ such that $\|x_p(s,\lambda)\| \leq M$, $\lambda \in \Pi_p$. Then we show that Π_p is closed relative to $(0, \mu_p^{(o)})$. Indeed let $\{\lambda_k\}$, $\lambda_k \in \Pi_p$, $k = 0,1,2,\cdots$ be a convergent sequence. Each function $x_p^k = x_p(s,\lambda_k)$ solves eq. (6.2) with $\lambda = \lambda_k$, i.e. $\lambda_k x_p^k = Kfx_p^k$, and $\|x_p^k\| \leq M$. Since K is compact there exists a subsequence $\{\lambda_{k_1}\}$ such that $\{Kfx_p^{k_1}\}$ converges in norm. We have that $Kfx_p^{k_1} = \lambda_{k_1} x_p^{k_1}$ however, whence $\{x_p^{k_1}\}$ converges in norm. (We may consider here that $\{\lambda_{k_1}\}$ is bounded away from 0; otherwise we should already be done.) Then $\bar{x}_p = \lim_{k_1\to\infty} x_p^{k_1}$ is a solution of eq. (6.2) with $\lambda = \bar{\lambda} = \lim_{k_1\to\infty} \lambda_{k_1}$, i.e. $\bar{\lambda}\,\bar{x}_p = Kf\bar{x}_p$, and by Theorem 6.1 $\bar{x}_p$ has exactly p nodes in $0 < s < 1$. Hence $\bar{\lambda} \in \Pi_p$, which shows that Π_p is closed. On the other hand Π_p must be open relative to $(0,\mu_p^{(o)})$ since, given $\bar{\lambda} \in \Pi_p$, $x_p(s,\bar{\lambda}) = \bar{x}_p$ exists such that $\bar{\lambda}\,\bar{x}_p = Kf\bar{x}_p$, $[\bar{\lambda}I - Kf'_{\bar{x}_p}]^{-1}$ exists as a bounded inverse by Theorem 7.1, and Theorem 2.3 indicates that there is a neighborhood $N_{\bar{\lambda}}$ of $\bar{\lambda}$ such that $x_p(s,\lambda)$ exists, $\lambda \in N_{\bar{\lambda}}$; i.e. $N_{\bar{\lambda}} \subset \Pi_p$. A set such as Π_p which is both open and closed relative to $(0,\mu_p^{(o)})$ must either be empty or be equal to $(0,\mu_p^{(o)})$. We have seen however that Π_p is not empty since it contains a left neighborhood of $\mu_p^{(o)}$. Hence $\Pi_p = (0,\mu_p^{(o)})$ under the assumption that $\|x_p(s,\lambda)\| \le M$, $\lambda \in \Pi_p$. But by Theorem 6.2, we must have $\mu_p < \lambda$ for $\lambda \in \Pi_p$, where μ_p is the p'th eigenvalue of linearized eq. (6.1) with $y(s) = x_p(s,\lambda)$. Now for functions $x(s) \in C(0,1)$ with $\|x\| \le M$, we have $f'_x(s,x) \ge f'_x(s,M)$ by the sublinearity assumption, (H-4a). Let μ_{pM} be the p'th eigenvalue of the operator $Kf'_M = \int_0^1 K(s,t)f'_x(t,M)\cdot dt$. Using the symmetrized operators of Theorem 6.2, which are such that Hf'_xH has the same spectrum as Kf'_x, we can write

$$\mu_{pM} = \min_{\{v_1,\cdots,v_{p-1}\}} \max_{u \perp v_1,\cdots,v_{p-1}} \frac{(Hf'_M Hu,u)}{(u,u)}$$

$$\leq \max_{u \perp w_1,\cdots w_{p-1}} \frac{(Hf'_M Hu,u)}{(u,u)} \leq \max_{u \perp w_1,\cdots,w_{p-1}} \frac{(Hf'_{x_p} Hu,u)}{(u,u)} = \mu_p,$$

where we have indicated here by $w_1,\cdots,w_{p-1}$ the first p-1 eigenelements of the operator $Hf'_{x_p}H$. Hence for $\lambda \in \Pi_p$ we necessarily have $0 < \mu_{pM} \leq \mu_p < \lambda$ under the assumption that $\|x_p(s,\lambda)\| \leq M$. This is a contradiction since we also proved that $\Pi_p = (0,\mu_p)$. Thus $x_p(s,\lambda)$ cannot remain bounded; there exists a number $\bar{\lambda} \geq 0$ such that as $\lambda \to \bar{\lambda}$, $\lambda > \bar{\lambda}$, $\lim\sup \|x_p(s,\lambda)\| = \infty$. In the superlinear case (H-4b) the argument is the same except that the set Π_p, where $x_p^{\pm}(s,\lambda)$ exists and is continuous, lies to the right of the primary bifurcation point $\mu_p^{(o)}$, and is a subset of the interval $(\mu_p^{(o)},\infty)$. This proves the theorem.

Now let us consider the linear eigenvalue problem

$$\gamma h(s) = \int_0^1 K(s,t)A(t)h(t)dt \tag{7.1}$$

formed with the function $A(s) \geq 0$ of the Hypothesis H-5 of Section 6. Problem (7.1) possesses the positive sequence of simple eigenvalues $\{\gamma_n\}$.

Finally, to prove the following result, we must strengthen H-5:

H-7: $|f(s,\beta)-A(s)\beta| \leq M_1$, $0 \leq \beta < \infty$, where M_1 is a constant.

Theorem 7.3: $\bar{\lambda} = \gamma_p$, where $\bar{\lambda} \geq 0$ is the number appearing in the last result; i.e. $\lim\sup_{\lambda \to \gamma_p} \|x_p^{\pm}(s,\lambda)\| = \infty$.

Proof: Let KA be the operator of eq. (7.1). By Theorem 7.2 there exists a sequence $\lambda_k \to \bar{\lambda}$ such that $\lim_{\lambda_k \to \bar{\lambda}} \|x_p(s,\lambda_k)\| = \infty$, and $x_p(s,\lambda_k)$ satisfies eq. (6.2) with $\lambda = \lambda_k$. We subtract the element KAx_p from each side of eq. (6.2) to obtain

$$(\lambda I-KA)x_p(s,\lambda_k) = \int_0^1 K(s,t)[f(t,x_p(t,\lambda_k))-A(t)x_p(t,\lambda_k)]dt,$$

whence, using the sup norm of C(0,1),

$$\|x_p(s,\lambda_k)\| \le \|(\lambda I-KA)^{-1}\|\cdot\|K\|\cdot\|f(s,x_p(s,\lambda_k))-A(s)x_p(s,\lambda_k)\|.$$

Then by H-7, $\|x_p(s,\lambda_k)\| \le \|(\lambda I-KA)^{-1}\|\cdot\|K\|\cdot M_1$. Thus $\lim_{\lambda_k \to \bar{\lambda}} \|x_p(s,\lambda_k)\| = \infty$ implies that $\bar{\lambda} \in \{\gamma_n\}$ where $\{\gamma_n\}$ are the eigenvalues of eq. (7.1).

Suppose now that $\bar{\lambda} = \gamma_m > 0$. We compare the functions h_m^∞ and $x_p(s,\lambda_k)$ where $h_m^\infty(s)$ is the normalized eigenfunction associated with γ_m.

$$h_m^\infty(s) - \frac{x_p(s,\lambda_k)}{\|x_p(s,\lambda_k)\|} = \frac{1}{\gamma_m}\int_0^1 K(s,t)A(t)h_m^\infty(t)dt - \frac{1}{\lambda_k\|x_p(s,\lambda_k)\|}\int_0^1 K(s,t)f(t,x_p(t,\lambda_k))dt$$

$$= \left(\frac{1}{\gamma_m} - \frac{1}{\lambda_k}\right)\int_0^1 K(s,t)A(t)h_m^\infty(t)dt + \frac{1}{\lambda_k}\int_0^1 K(s,t)A(t)\left[h_m^\infty(t) - \frac{x_p(t,\lambda_k)}{\|x_p(t,\lambda_k)\|}\right]dt$$

$$+ \frac{1}{\lambda_k\|x_p(s,\lambda_k)\|}\int_0^1 K(s,t)[A(t)x_p(t,\lambda_k)-f(t,x_p(t,\lambda_k))]dt.$$

Now $\left|\frac{1}{\gamma_m} - \frac{1}{\lambda_k}\right| \cdot \|KA\| \to 0$ as $\lambda_k \to \gamma_m$, and likewise

8. The Example of Section 1, Reconsidered.

Having developed some methods for treating more general cases, let us now reconsider the example of section 1, namely eq. (1.1). This equation now appears to be an eigenvalue problem for a superlinear Hammerstein operator of odd type. In fact we find that hypotheses H-1 through H-3, H-4b, and H-5 with $A(s) = A = \infty$, are satisfied. The second rank kernel does not impress us as being an oscillation kernel in that it is possible for it to assume negative values, but in a simple example we can live with whatever deficient properties a specific kernel does have, if it has any, and find out where we are led.

Accordingly let us begin by treating eq. (1.1) in a fashion reminiscent of Section 2. Namely, let h,δ be increments added to φ_o,λ_o respectively, where we assume that (φ_o,λ_o) represents a pair which satisfies eq. (1.1). We have then

$$\lambda_o\varphi_o + \lambda_o h + \delta\varphi_o + \delta h = \frac{2}{\pi}\int_0^\pi [a \sin s \sin t + b \sin 2s \sin 2t] \cdot \{\varphi_o + h + \varphi_o^3 + 3\varphi_o^2 h + 3\varphi_o h^2 + h^3\} dt. \tag{8.1}$$

The fact that (φ_o,λ_o) solves eq. (1.1) allows some cancellations in eq. (8.1); after rearrangement we get

$$\left\{\lambda_o I - \frac{2}{\pi}\int_0^\pi [a \sin s \sin t + b \sin 2s \sin 2t][1 + 3\varphi_o^2(t)]\cdot dt\right\} h =$$

$$= -\delta\varphi_o - \delta h + \frac{2}{\pi}\int_0^\pi [a \sin s \sin t + b \sin 2s \sin 2t][3\varphi_o(t)h^2(t) + h^3(t)]dt = F_\delta(h). \tag{8.2}$$

At the trivial solution $\varphi_o(s) \equiv 0$ we have the linearization eq. (1.5) with two eigenvalues a,b, with a > b as assumed, to which are associated the eigenspaces spanned respectively by sin s, sin 2s. We thus have two primary bifurcation points, a,b, where the operator on the left in eq. (8.2) has no inverse. Corresponding to eq. (3.2) we have the following equation to be solved for h at the bifurcation point $\lambda = a$, (with $\delta = \lambda - a$):

$$h = \overline{M}F_\delta(h) + \xi \sin s = M(I-E)\left\{-\delta h + \frac{2}{\pi}\int_0^\pi [a \sin s \sin t + b \sin 2s \sin 2t] h^3(t)dt\right\} + \xi \sin s$$

where here E is the orthogonal projection on the null space spanned by sin s, and M is the pseudo inverse. This gives

$$h = M\left\{-\delta(I-E)h + \frac{2}{\pi}\int_0^\pi b \sin 2s \sin 2t\, h^3(t)dt\right\} + \xi \sin s. \tag{8.3}$$

Putting $h_o = \xi \sin s$ in an iteration process, we find that the integral in eq. (8.3) vanishes, so that $h_1 = \xi \sin s$. Likewise every succeeding iterate is equal to $\xi \sin s$, and therefore $h = V_\delta(\xi \sin s) = \xi \sin s$. Then the bifurcation equation, eq. (3.4), becomes

$$EF_\delta(\xi \sin s) = E\left\{-\delta\xi \sin s + \frac{2}{\pi}\int_0^\pi [a \sin s \sin t + b \sin 2s \sin 2t]\xi^3 \sin^3 t\, dt\right.$$

$$= \{-\delta\xi + \tfrac{3}{4} a\xi^3\} \sin s = 0. \tag{8.4}$$

Eq. (8.4) has the trivial solution $\xi = 0$ and the nontrivial solution $\xi = \pm\frac{2}{\sqrt{3}}\sqrt{\frac{\delta}{a}} = \pm\frac{2}{\sqrt{3}}\sqrt{\frac{\lambda}{a} - 1}$, and the first branch solution, obtained by substituting this ξ into eq. (8.3), is $h(s) = \pm\frac{2}{\sqrt{3}}\sqrt{\frac{\lambda}{a} - 1}\,\sin s$.

Again, we write eq. (3.2) for the bifurcation point at $\lambda = b$, with $\delta = \lambda - b$:

$$h = M(I-E)\left\{-\delta h + \frac{2}{\pi}\int_0^{\pi}[a \sin s \sin t + b \sin 2s \sin 2t]h^3(t)dt\right\} + \xi \sin 2s$$

where E is now the orthogonal projection onto the null space spanned by sin 2s, and $\delta = \lambda - b$. This gives

$$h = M\left\{-\delta(I-E)h + \frac{2}{\pi}\int_0^{\pi} a \sin s \sin t\, h^3(t)dt\right\} + \xi \sin 2s. \qquad (8.5)$$

Starting with the first iterate $h_0 = \xi \sin 2s$ and substituting this on the right in eq. (8.5), we again have the integral vanishing, whence $h_1 = \xi \sin 2s$. We can then see that $h_n = \xi \sin 2s$ also, so that $h = V_\delta(\xi \sin 2s) = \xi \sin 2s$. The bifurcation equation is now written

$$EF_\delta(\xi \sin 2s) = E\left\{-\delta\xi \sin 2s + \frac{2}{\pi}\int_0^{\pi}[a \sin s \sin t + b \sin 2s \sin 2t]\xi^3\sin^3 2t\, dt\right.$$

$$= \{-\delta\xi + \tfrac{3}{4}b\xi^3\} \sin 2s = 0. \qquad (8.6)$$

Eq. (8.6) has the trivial solution $\xi = 0$ and the nontrivial solution $\xi = \pm \frac{2}{\sqrt{3}}\sqrt{\frac{\delta}{b}} = \pm \frac{2}{\sqrt{3}}\sqrt{\frac{\lambda}{b} - 1}$. We solve eq. (8.5), with this solution of eq. (8.6), to get $h(s) = \pm \frac{2}{\sqrt{3}}\sqrt{\frac{\lambda}{b} - 1} \sin 2s$ for the second branch of eigensolutions.

These branches of eigensolutions are exactly the same as those obtained in Section 1 by more elementary means. Also since the expansion (3.5) is trivial in this case, the expressions $\varphi_1(s,\lambda) = \varphi_0 + h = \pm \frac{2}{\sqrt{3}}\sqrt{\frac{\lambda}{a} - 1} \sin s$ and

$\varphi_2(s,\lambda) = \varphi_0 + h = \pm \frac{2}{\sqrt{3}}\sqrt{\frac{\lambda}{b} - 1}$ sin 2s are valid in the large. There is no need for the process of Theorems 2.3 and 2.4. Of course one could follow the steps. The uniqueness property of Theorem 2.3 would yield no other result.

With this example in Section 1 however, we had secondary bifurcation on the 1st branch if $\frac{b}{a} > \frac{1}{2}$. Here, we study this possibility by learning how the eigenvalue μ_2 of the linearization

$$\mu h(s) = \frac{2}{\pi}\int_0^{\pi} [a \sin s \sin t + b \sin 2s \sin 2t][1+3\varphi_1^2(t,\lambda)]h(t)dt \tag{8.7}$$

behaves as the 1st branch $\varphi_1(s,\lambda)$ evolves. The eigenvalue μ_1 does not bother us since $\mu_1 = a+3(\lambda-a) = \lambda+2(\lambda-a) > \lambda$. Of course this is what Theorem 6.2 would tell us. For μ_2 we have the expression $\mu_2 = -b + \frac{2b}{a}\lambda$. Secondary bifurcation of the branch $\varphi_1(s,\lambda)$ occurs if ever $\mu_2 = \lambda$; this does happen if $\frac{2b}{a} > 1$ but cannot happen if $\frac{2b}{a} \le 1$. Hence we get the same condition as in Section 1. The secondary bifurcation in this example occurs then at $\lambda_{sb} = \frac{ab}{2b-a}$ with the solution $\varphi_{sb}(s) = \pm \frac{2}{\sqrt{3}}\sqrt{\frac{a-b}{2b-a}}$ sin s.

There is of course the question of the bifurcation analysis at the secondary bifurcation point $(\varphi_{sb},\lambda_{sb})$. In Section 1 it was found, using direct elementary methods, that the two sub-branches or twigs split away from the main branch $\varphi_1(s,\lambda)$ at this point and evolve to the right. When it comes to repeating this bifurcation analysis by use of the bifurcation equation, eq. (4.2), we find that difficulties arise. When we compute the coefficients using (5.3), we find that a_1 vanishes; the nonvanishing of a_1 is essential in the application of the Newton Polygon method as discussed by J. Dieudonné [ref. 8, p. 4]. In treating bifurcation at the origin as in Section 4, we were able to handle a case where a_1 vanished since there it was clear that $\Phi_1(\delta,\xi_1)$ in eq. (4.2) vanished also. In this example where we have secondary bifurcation at $(\varphi_{sb},\lambda_{sb})$ we have yet to learn how to work out the sub-branches using the bifurcation equation.*

The vanishing of a_1 in eq. (4.2) at a secondary bifurcation point as above is a peculiarity of Hammerstein operators $\int_0^1 K(s,t)f(t,x(t))dt = Kfx$ for which $K(1-s,1-t) = K(s,t)$ and $f(1-s,x) = f(s,x)$. The example of section 1 is of this type. More general examples lead to the nonvanishing of a_1 in eq. (4.2) at a secondary bifurcation point, whence the Newton Polygon method is applicable to eq. (4.2) as it stands. It should be noted however that in a case of nonvanishing a_1 in eq. (4.2) one usually has a changing of direction of evolution of the branch of eigensolutions at the secondary bifurcation point, rather than a formation of sub-branches as in the problem of Section 1. Some writers prefer to call such a point a limit point of the branch of eigensolutions, thus preserving the term "secondary bifurcation" for the more intuitive idea of the splitting of a branch. In any case, however, the bifurcation analysis must be used in general.

We can compare and assess the two conditions against secondary bifurcation given in Section 6, namely (6.9b) and (6.15b) respectively. We found in Section 1 that a necessary and sufficient condition against secondary bifurcation in the example was that $\frac{b}{a} \leq \frac{1}{2}$. How do conditions (6.9b) and (6.15b) compare with this?

---

*See Appendix however.

With respect to condition (6.9b), the quantity on the left, namely $\frac{\varphi(1+3\varphi^2)}{\varphi+\varphi^3}$, varies between 1 and 3. For the condition to be satisfied Λ_2 must be no higher than $\frac{1}{3}$. Now in the present example, Λ_2 can be given a most refined definition. In connection with eq. (8.7) we saw that the two eigenvalues, $\mu_1(K,1+3\varphi^2) = 3\lambda-2a$ and $\mu_2(K,1+3\varphi^2) = -b+\frac{2b}{a}\lambda$ evolved as the first branch $\varphi_1(s,\lambda)$ evolved. We know these expressions for μ_1 and μ_2 only because we know a priori, independently of μ_1 and μ_2, the expression for $\varphi_1(s,\lambda)$ in this example. Hence we can compute the maximum of $\frac{\mu_2(K,1+3\varphi^2)}{\mu_1(K,1+3\varphi^2)}$ over this known branch only, rather than over the positive cone. The maximum of the ratio is $\frac{b}{a}$ and is assumed at $\lambda = a$, i.e. at the origin $\varphi \equiv 0$. This allows interpretation of condition (6.9b) in terms of eigenvalues of the linearization at the origin, namely a and b. Condition (6.9b) therefore requires $\frac{b}{a} \leq \frac{1}{3}$ as a condition for no secondary bifurcation in this example. Hence condition (6.9b) while being a sufficient condition, is far from being a necessary condition for no secondary bifurcation.

Condition (6.15b) on the other hand requires that

$$\frac{1+3\varphi_1^2}{1+\varphi_1^2} < \frac{\mu_1(K,1+3\varphi_1^2)}{\mu_2(K,1+3\varphi_1^2)}$$

as a condition against secondary bifurcation. With $\varphi_1 = \varphi_1(s,\lambda) = \pm\frac{2}{\sqrt{3}}\sqrt{\frac{\lambda}{a}-1}\,\sin s$ this means that the condition is satisfied if

$$\frac{1+4\left(\frac{\lambda}{a}-1\right)}{1+\frac{4}{3}\left(\frac{\lambda}{a}-1\right)} < \frac{3\lambda-2a}{2\frac{b}{a}\lambda-b} .$$

Indeed the latter is satisfied for $a \leq \lambda \leq \infty$ provided $\frac{b}{a} \leq \frac{1}{2}$. Hence in the example, condition (6.15b) stacks up quite well as a condition against secondary bifurcation.

9. A Two-Point Boundary Value Problem.

In eq. (6.2), if we let $K(s,t)$ be the Green's function

$$G(s,t) = \begin{cases} \frac{\pi-t}{\pi} s, & s < t \\ \frac{\pi-s}{\pi} t, & s > t \end{cases} \tag{9.1}$$

$$= \frac{2}{\pi} \sum_{n=1}^{\infty} \frac{1}{n^2} \sin ns \sin nt, \quad 0 \leq s, t \leq \pi,$$

we have an integral equation which is equivalent to the following two-point boundary value problem (we have changed the upper limit of integration in eq. (6.2) to π):

$$x_{ss} + \lambda^{-1} f(s,x(s)) = 0$$
$$x(o) = 0 \qquad x(\pi) = 0. \tag{9.2}$$

The question arises of what alternative methods there may be for studying branches of eigenfunctions and their bifurcations for the equivalent two-point boundary-value problem. In this connection please see papers by the author [ref. 17] and C. V. Coffman [ref. 6]. Coffman gives certain interesting conditions against secondary bifurcation in problems such as eq. (9.2). In particular there is no secondary bifurcation of the branches of eigenfunctions in the autonomous case, $f(s,x(s)) \equiv f(x(s))$. We now study the autonomous case to see how the absence of secondary bifurcation on any of the branches of eigenfunctions can be related to our considerations involving Hammerstein's equation, eq. (6.2).

We take a definite case; namely let us deal with the Hammerstein Equation

$$\lambda\varphi(s) = \int_0^{\pi} G(s,t)[\varphi(t)+\varphi^3(t)]dt \tag{9.3}$$

and the equivalent boundary value problem

$$\varphi_{ss} + \lambda^{-1}[\varphi+\varphi^3] = 0$$

$$\varphi(0) = 0 \qquad \varphi(\pi) = 0. \tag{9.4}$$

The first step in our solution of problem (9.4) is to let $\xi = \lambda^{-\frac{1}{2}}s$. Then $\varphi_s = \lambda^{-\frac{1}{2}}\varphi_\xi$ and $\varphi_{ss} = \lambda^{-1}\varphi_{\xi\xi}$, whence the problem takes the form

$$\varphi_{\xi\xi} + [\varphi+\varphi^3] = 0$$

$$\varphi(0) = 0 \qquad \varphi(\lambda^{-\frac{1}{2}}\pi) = 0. \tag{9.5}$$

This sets the stage for using the initial value problem

$$\varphi_{\xi\xi} + [\varphi+\varphi^3] = 0$$

$$\varphi(o) = 0, \qquad \varphi_\xi(o) = c > 0 \tag{9.6}$$

as a tool in solution of boundary value problems (9.5) and (9.4).

Problem (9.6) has the first integral

$$(\varphi_\xi)^2 + [\varphi^2 + \tfrac{1}{2}\varphi^4] = c^2$$

which defines a closed trajectory in the φ,φ_ξ phase plane (see Fig. 9.1). Then $\varphi_\xi = \pm\sqrt{c^2-\varphi^2-\frac{1}{2}\varphi^4}$, and in the first quadrant we have

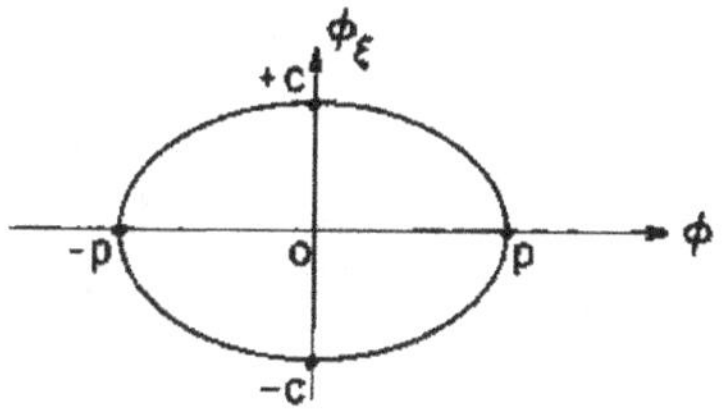

FIG. 9.1.

$$\xi = \int_0^\varphi \frac{d\bar{\varphi}}{\sqrt{c^2-\bar{\varphi}^2-\frac{1}{2}\bar{\varphi}^4}}. \tag{9.7}$$

We factor the denominator in the integral of eq. (9.7):

$$\tfrac{1}{2}\bar{\varphi}^4 + \bar{\varphi}^2 - c^2 = \tfrac{1}{2}(\bar{\varphi}^2-[-1+\sqrt{1+2c^2}])(\bar{\varphi}^2-[-1-\sqrt{1+2c^2}]).$$

Defining $p^2 = -1 + \sqrt{1+2c^2}$ and $q^2 = 1 + \sqrt{1+2c^2}$, we write (9.7) as follows:

$$\xi = \sqrt{2} \int_0^{\varphi} \frac{d\bar{\varphi}}{[p^2-\bar{\varphi}^2][q^2+\bar{\varphi}^2]}, \qquad 0 \le \varphi \le p,$$

$$= \sqrt{2}\ \frac{1}{\sqrt{p^2+q^2}}\ sn^{-1}(\sin\psi, k)$$

where sn is the Jacobi Elliptic function of the 1st kind [ref. 5, p. 50 #214.00] and we have the following definitions:

$$k^2 = \frac{p^2}{p^2+q^2} = \frac{-1+\sqrt{1+2c^2}}{2\sqrt{1+2c^2}} \quad \text{and} \quad \psi = \sin^{-1}\sqrt{\frac{\varphi^2(p^2+q^2)}{p^2(\varphi^2+q^2)}} .$$

Inverting we get $\sin\psi = sn\left(\sqrt{\frac{p^2+q^2}{2}}\ \xi, k\right) = \sqrt{\frac{\varphi^2(p^2+q^2)}{p^2(\varphi^2+q^2)}}$. Solving we have $\varphi^2(p^2+q^2) = p^2(\varphi^2+q^2)\,sn^2\left(\sqrt{\frac{p^2+q^2}{2}}\ \xi, k\right)$

or $\varphi^2\left[p^2+q^2-p^2 sn^2\left(\sqrt{\frac{p^2+q^2}{2}}\ \xi, k\right)\right] = p^2q^2 sn^2\left(\sqrt{\frac{p^2+q^2}{2}}\ \xi, k\right)$.

Thus $$\varphi(\xi) = \frac{pq\ sn\left(\sqrt{\frac{p^2+q^2}{2}}\ \xi, k\right)}{\sqrt{p^2+q^2-p^2\ sn^2\left(\sqrt{\frac{p^2+q^2}{2}}\ \xi, k\right)}}$$

$$= \frac{\sqrt{2}\ c\ sn\left(\sqrt[4]{1+2c^2}\ \xi, k\right)}{\sqrt{2\sqrt{1+2c^2} - \left[-1+\sqrt{1+2c^2}\right] sn\left(\sqrt[4]{1+2c^2}\ \xi, k\right)}}$$

is the solution of problem (9.6) for $0 < \xi < \int_0^p \frac{d\bar{\varphi}}{\sqrt{c^2-\bar{\varphi}^2-\frac{1}{2}\bar{\varphi}^4}}$, and indeed for all subsequent values of ξ.

The elliptic function of the first kind $\mathrm{sn}(\tau,k)$ has the real period $4K$ (and of course also a complex period) where

$$K(k) = \int_0^{\frac{\pi}{2}} \frac{d\theta}{\sqrt{1-k^2\sin^2\theta}} = \int_0^1 \frac{dt}{\sqrt{(1-t^2)(1-k^2t^2)}}$$

is monotone increasing in k, [ref. 5, p. 19]. Hence if we seek to solve the boundary value problem (9.5), we are interested in matching the zeros of $\mathrm{sn}\left(\sqrt[4]{1+2c^2}\ \xi, k\right)$, which occur at $\xi = \frac{2nK(k)}{\sqrt[4]{1+2c^2}}$ $n = 0,1,2,\cdots$ with the value $\xi = \lambda^{-\frac{1}{2}}\pi$. Let the first such zero be ξ_1; then we have

$$\sqrt[4]{1+2c^2}\ \xi_1 = 2K(k)$$

and

$$\xi_1(c) = \frac{2K(k)}{\sqrt[4]{1+2c^2}} = \frac{2}{\sqrt[4]{1+2c^2}} \int_0^1 \frac{dt}{\sqrt{(1-t^2)(1-k^2t^2)}}$$

where we remember that $k = k(c) = \sqrt{\frac{-1+\sqrt{1+2c^2}}{2\sqrt{1+2c^2}}}$. We have $\frac{\pi}{2} \le K(k) < \frac{\pi}{\sqrt{2}}$,

so that $\xi_1(c) \to 0$ as $c \to \infty$ and $\xi_1(c) \to \pi$ as $c \to 0$. It can also be verified that $\xi_1(c)$ is monotone decreasing, $c > 0$. Of interest is the solution c_1 of the equation $\xi_1(c) = \lambda^{-\frac{1}{2}}\pi$, (see Fig. 9.2). With c_1 thus defined, we can write

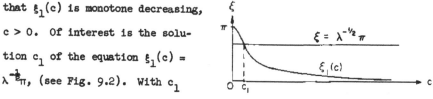

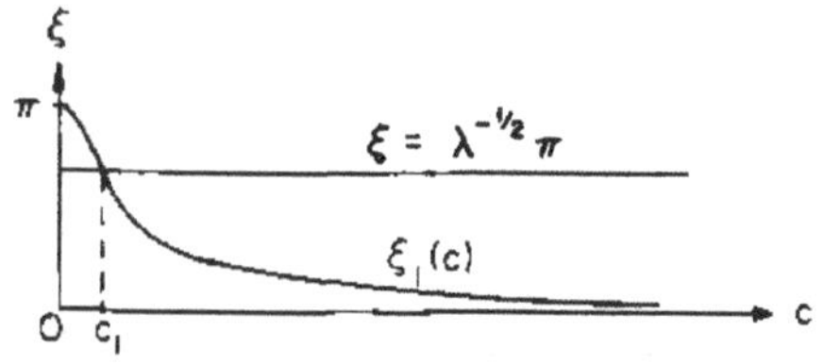

FIG. 9.2.

$$\lambda = \frac{\pi^2}{\xi_1^{\,2}(c_1)} = \frac{\pi^2\sqrt{1+2c_1^{\,2}}}{4K^2(k(c_1))} \tag{9.8}$$

which goes to 1 as $c_1 \to 0$, and to ∞ as $c_1 \to \infty$.

"Thus c_1 is the initial slope of an eigenfunction of problem (9.5), and eq. (9.8) tells how the eigenvalue λ varies as we vary the initial slope. Thus eq. (9.8), together with problem (9.6), defines a branch of eigenfunctions of boundary value problem (9.5) which is parameterized by the initial slope $c_1 > 0$. It is the zero'th branch of eigenfunctions which bifurcates from the trivial solution $\varphi \equiv 0$ at the zero'th bifurcation point $\lambda = 1$, which is the zero'th eigenvalue of the linearized problem at the origin:

$$\begin{aligned} &\varphi_{\xi\xi} + \varphi = 0 \\ &\varphi(o) = 0 \qquad\qquad \varphi(\lambda^{-\frac{1}{2}}\pi) = 0. \end{aligned} \tag{9.9}$$

Linear problem (9.9) has simple eigenvalues at $\lambda = \frac{1}{n^2}$, $n = 1,2,\cdots$.

In a similar fashion we match the n'th zero ξ_n of the solution of problem (9.6) with $\lambda^{-\frac{1}{2}}\pi$. We have

$$\sqrt[4]{1+2c^2}\;\xi_n = 2nK(k)$$

and so get the expression

$$\xi_n(c) = \frac{2nK(k)}{\sqrt[4]{1+2c^2}} = n\xi_1(c).$$

Solution of the equation $\xi_n(c) = \lambda^{-\frac{1}{2}}\pi$ yields a value c_n. Then we have

$$\lambda = \frac{\pi^2}{\xi_n^2(c_n)} = \frac{\pi^2\sqrt{1+2c_n^2}}{(2n)^2K^2(k(c_n))} \to \frac{1}{n^2} \quad \text{as} \quad c_n \to 0 \tag{9.10}$$

and $\lambda \to \infty$ as $c_n \to \infty$.

Thus is yielded the (n-1)th branch of eigenfunctions of problem (9.5), that branch with n-1 interior zeros. It is parameterized by the initial slope c_n; eq. (9.10) gives the eigenvalue while problem (9.6) with $c = c_n$ gives the eigenfunctions. The primary bifurcation from $\varphi \equiv 0$ is at $\lambda = \frac{1}{n^2}$.

For future reference, we note that the n'th maximum of the solution of problem (9.6) (or the n'th zero of the function ψ below) occurs at

$$\eta_n(c) = \frac{(2n-1)K(k(c))}{4\sqrt{1+2c^2}} . \tag{9.11}$$

Now let us consider again the solution of problem (9.6) which forms a trajectory $\psi^2+[\varphi^2+\frac{1}{2}\varphi^4] = c^2$ in the φ,ψ phase plane (see Fig. 9.1). Here we have set $\psi = \varphi_\xi$. Let ξ_ν = successive zeros of φ, $\nu = 0,1,2,\cdots$

η_ν = successive zeros of ψ, $\nu = 1,2,3,\cdots$

where we label $\xi_0 = 0$ arbitrarily. By inspection of Fig. 9.1 we note that

$$\operatorname{sgn} \varphi(\xi,c) = (-1)^\nu, \qquad \xi_\nu < \xi < \xi_{\nu+1}$$

$$\operatorname{sgn} \psi(\xi,c) = (-1)^\nu, \qquad \eta_\nu < \xi < \eta_{\nu+1}.$$

We consider also the linearized initial value problem

$$h_{\xi\xi} + [1+3\varphi^2]h = 0$$
$$h(o) = 0 \qquad h_\xi(o) = 1. \tag{9.12}$$

Problem (9.12) has a trajectory which also revolves around the origin of an h,k phase plane, where $k = h_\xi$. In Fig. 9.3 we superimpose the two phase planes. Define α_ν = the successive zeros of h, ν = 0,1,2,···. By inspection we have

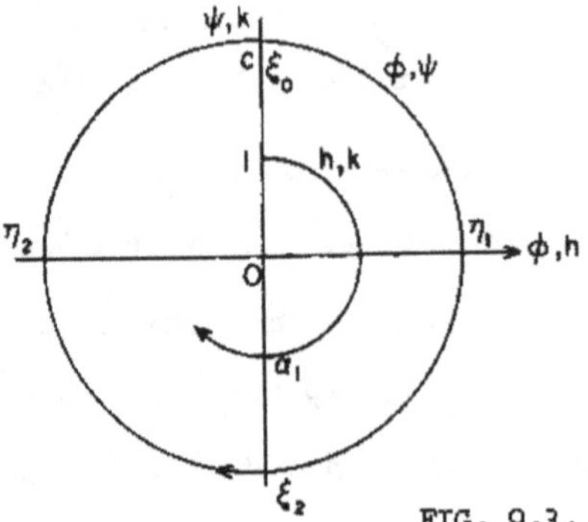

FIG. 9.3.

$$\operatorname{sgn} h(\xi,c) = (-1)^\nu, \quad \alpha_\nu < \xi < \alpha_{\nu+1}.$$

If we multiply the differential equation $h_{\xi\xi} + [1+3\varphi^2]h = 0$ through by φ, multiply the differential equation $\varphi_{\xi\xi} + [\varphi+\varphi^3] = 0$ through by h, and subtract the latter from the former, we get

$$[\varphi h_{\xi\xi} - h\varphi_{\xi\xi}] + [(1+3\varphi^2) - (1+\varphi^2)]h\varphi = 0.$$

Integration from ξ_ν to $\xi_{\nu+1}$ gives

$$-h(\xi_{\nu+1},c)\psi(\xi_{\nu+1},c) + h(\xi_\nu,c)\psi(\xi_\nu,c) + 2\int_{\xi_\nu}^{\xi_{\nu+1}} \varphi^2 \cdot h\varphi d\xi = 0. \qquad (9.13)$$

Lemma 9.1: $\alpha_\nu < \xi_\nu$, $\nu = 1,2,3,\cdots$; in other words the h,k trajectory leads the φ,ψ trajectory in Fig. 9.3.

Proof: We employ induction. Assume the lemma is true for $\nu = 1,2,\cdots,m$, but not for m+1. Then we have $\alpha_m < \xi_m < \xi_{m+1} \le \alpha_{m+1}$. The integrand in eq. (9.13), (in which we put $\nu = m$), is positive since h and φ have the same sign between ξ_m and ξ_{m+1} under our assumption, and since $\operatorname{sgn} h(\xi_m,c) = (-1)^m$, and $\operatorname{sgn} \psi(\xi_m,c) = (-1)^m$, we also have $h(\xi_m,c)\psi(\xi_m,c) > 0$. Thus by eq. (9.13) we should have $h(\xi_{m+1},c)\psi(\xi_{m+1},c) > 0$. The latter must be false however since either sgn $h(\xi_{m+1},c) = (-1)^m$ or $h(\xi_{m+1},c) = 0$, by our assumption, but sgn $\psi(\xi_{m+1},c) = (-1)^{m+1}$. This contradiction shows that $\alpha_{m+1} < \xi_{m+1}$, and proves the lemma.

A proof that $\varphi_c = \frac{\partial}{\partial c}\varphi(\xi,c) = h$ and $\psi_c = \frac{\partial}{\partial c}\psi(\xi,c) = k$ can be patterned after a very similar proof in a published paper of the author; [ref. 17; p. 132, Lemma 1].

Lemma 9.2: $\xi_\nu < \alpha_{\nu+1}$, and $\eta_{\nu+1} < \alpha_{\nu+1}$, $\nu = 0,1,2,\cdots$; in other words, the lead of the (h,k) trajectory over the (φ,ψ) trajectory in Fig. 9.3 is less than 90°.

Proof: The first statement follows from the second since by inspection of Fig. 9.3, it is clear that $\xi_\nu < \eta_{\nu+1}$, $\nu = 0,1,2,\cdots$. The second statement can be proven by showing that sgn $h(\eta_\nu,c) =$ sgn $\varphi(\eta_\nu,c)$, $\nu = 1,2,3,\cdots$ (with reference to Lemma 9.1). From the expression of the solution of problem (9.6) in terms of elliptic functions, we have $\varphi(\eta_\nu,c) = (-1)^{\nu-1}p = (-1)^{\nu-1}\sqrt{\sqrt{1+2c^2}-1}$. Now $\frac{d}{dc}\varphi(y_\nu,c) = \varphi_\xi(\eta_\nu(c),c)\frac{d\eta_\nu}{dc} + h(\eta_\nu,c)$. But $\varphi_\xi(\eta_\nu(c),c) = \psi(\eta_\nu,c) = 0$. Therefore $\frac{d}{dc}\varphi(\eta_\nu,c) = h(\eta_\nu,c) = (-1)^{\nu-1}\frac{d}{dc}\sqrt{\sqrt{1+2c^2}-1}$

$$= (-1)^{\nu-1}\frac{1}{\sqrt{\sqrt{1+2c^2}-1}}\cdot\frac{c}{\sqrt{1+2c^2}}.$$ Also we have sgn $\varphi(\eta_\nu,c) = (-1)^{\nu-1}$.

Therefore sgn $h(\eta_\nu,c) =$ sgn $\varphi(\eta_\nu,c)$, $\nu = 1,2,3,\cdots$, which proves the lemma.

Theorem 9.3: There is no secondary bifurcation on any branch of eigenfunctions of problem (9.4).

Proof: Solutions of problem (9.5), equivalent to problem (9.4), are given by solutions of problem (9.6) with $c = c_n > 0$, where c_n solves the equation

$\xi_n(c) = \lambda^{-\frac{1}{2}}\pi$. By eq. (9.10) we had $\lambda = \dfrac{\pi^2}{\xi_n^2(c_n)}$ as the n'th branch eigenvalue of problem (9.5) expressed as a function of the initial slope c_n of the associated eigenfunction. In exactly the same way, the discrete eigenvalues $\{\mu_\nu\}$ of the linearized boundary value problem

$$\begin{aligned} & h_{\xi\xi} + (1+3\varphi_n^2)h = 0 \\ & h(o) = 0 \qquad\qquad h(\lambda^{-\frac{1}{2}}\pi) = 0 \end{aligned} \tag{9.14}$$

are obtained by matching the zeros α_ν of the solution of the initial value problem (9.12) with the value $\lambda^{-\frac{1}{2}}\pi$; thus $\mu_\nu = \dfrac{\pi^2}{\alpha_\nu^2(c_n)}$. By Lemmas 9.1 and 9.2 we can then make the following comparison:

$$\mu_{n+1}(c_n) = \frac{\pi^2}{\alpha_{n+1}^2(c_n)} < \frac{\pi^2}{\xi_n^2(c_n)} = \lambda < \frac{\pi^2}{\alpha_n^2(c_n)} = \mu_n(c_n)$$

$$0 \le c_n < \infty.$$

Hence nowhere on the n'th branch of eigenfunctions, for any n, do we have $\lambda = \mu_{n+1}$ or $\lambda = \mu_n$. Since the eigenvalues μ_n, μ_{n+1} of problem (9.14) are necessarily simple, we can conclude that there is no secondary bifurcation on any branch of eigenfunctions of problem (9.4). This concludes the proof.

In Section 6 we were concerned about the possibility $\mu_{n+1} = \lambda$, and devised two conditions against it. It is interesting to note here that in the case of problem (9.4), which of course is completely equivalent to the Hammerstein equation (9.3), we can say that $\mu_{n+1} < \lambda$ with a margin. Indeed, by Lemma 9.2 we have

$$\mu_{n+1}(c_n) = \frac{\pi^2}{\alpha_{n+1}^2(c_n)} < \frac{\pi^2}{\eta_{n+1}^2(c_n)} < \frac{\pi^2}{\xi_n^2(c_n)} = \lambda(c_n);$$

$$0 \le c_n < \infty.$$

Thus μ_{n+1} stays bounded away from λ on the n'th branch.

The following eigenvalue problem for a Hammerstein operator:

$$\lambda\varphi(s) = \int_0^{\pi} K(s,t)[\varphi(t)+\varphi^3(t)]dt \tag{9.15}$$

where we define the kernel as follows (with positive constants $\{\mu_n^{(o)}\}$ so chosen that the resulting operator is compact):

$$K(s,t) = \frac{2}{\pi} \sum_{n=1}^{\infty} \mu_n^{(o)} \sin ns \sin nt$$

is a generalization on the eigenvalue problem of Section 1. In Section 1 we merely chose the first two terms of this kernel. In the present section we have discovered that there indeed do exist sets of constants $\{\mu_n^{(o)}\}$ such that the eigenfunction branches of eq. (9.15) undergo no secondary bifurcation; namely $\mu_n^{(o)} = \frac{1}{n^2}$. Hence if some condition on the primary bifurcation points $\{\mu_n^{(o)}\}$ of problem (9.15) could be devised assuring against secondary bifurcation (the condition of Section 1 was that $\frac{\mu_1^{(o)}}{\mu_o^{(o)}} < \frac{1}{2}$), the constants $\mu_n^{(o)} = \frac{1}{n^2}$ would presumably have to satisfy it. Such a condition would not be vacuous therefore, and would be of a form much more convenient to apply than than those of section 6.

10. Summary; Collection of Hypotheses; Unsettled Questions.

To summarize, we state the assumptions made about the operator $T(x)$, and the conclusions, in a progressive fashion.

In section 2 it is assumed that $T(x)$ is a generally nonlinear continuous, Fréchet differentiable transformation, with continuous Fréchet derivative, defined on a Banach Space X with real scalar field, and mapping X into itself with $T(\theta) = \theta$, where θ is the null element of X. The Fréchet derivative of $T(x)$ is a linear operator $T'(x)$ defined for each parameter $x \in X$, and mapping X into itself. Let (x_0,λ_0) be an ordinary point for $T(x)$, i.e. the transformation $\lambda_0 I-T'(x_0)$ has a bounded inverse, and suppose also that (x_0,λ_0) solves the nonlinear equation (2.1), i.e., $\lambda_0 x_0 = T(x_0)+f$ where $f \in X$ is a fixed element. Then there exists in $X \times R$ a unique maximal sheet of solutions of eq. (2.1), namely (x,λ), which forms a locus of points in X parameterized by λ and denoted by $x(\lambda)$, (Theorem 2.4). This sheet is the only locus of solutions passing through the ordinary point (x_0,λ_0), and the only finite boundary points it can have in $X \times R$ are exceptional points, i.e., points (x_B,λ_B) such that $\lambda_B I-T'(x_B)$ does not have a bounded inverse.

Section 3 takes up the study of solutions of eq. (2.1) in a neighborhood of an exceptional point (x_0,λ_0). It is assumed here that $T'(x_0)$ is compact (statement H-1); such is the case if the nonlinear operator $T(x)$ is completely continuous on X. Let $\overline{M}$ be the extended pseudo-inverse of $\lambda_0 I-T'(x_0)$, (see Lemma 3.1). Let $h = V_\delta(u)$ be the unique solution of the equation $h = \overline{M}F_\delta(h)+u$ for $|\delta| < c$, where c is a constant of Theorem 3.2, and where $F_\delta(h) = - \delta x_0 - \delta h + R_1(x_0,h)$, $R_1(x_0,h)$ being the first order remainder in the Taylor expansion of $T(x)$ around x_0. Then the condition that $h = V_\delta(u)$ be such that $(x_0+h, \lambda_0+\delta)$ solves eq. (2.1) is that $(I-E_R)F_\delta(V_\delta(u)) = 0$, where E_R is the projection on the range of $\lambda_0 I-T'(x_0)$ along the null space of $\lambda_0 I-T'(x_0)$. Conversely if $(x_0+h, \lambda_0+\delta)$ satisfies eq. (2.1), then (h,δ) also satisfies $h = \tilde{M}F_\delta(h)+u$ where u conforms to the condition $(I-E_R)F_\delta(V_\delta(u)) = 0$. This bifurcation equation, namely $(I-E_R)F_\delta(V_\delta(u)) = 0$, is solved in the finite dimensional null space $\eta_1(x_0)$ and hence can be represented as a finite number of scalar equations in an equal finite number of unknowns. It was necessary to assume $T(x)$ to be three times Fréchet differentiable for this. The 1:1 correspondence between the solutions of the bifurcation equation and solutions of eq. (2.1) in a neighborhood of an exceptional point (x_0,λ_0) enables the study of solution multiplicity of eq. (2.1) by means of the actual solution multiplicity of an algebraic system, viz., the bifurcation equation. A concrete representation of this algebraic system in terms of assumed bases for the null spaces $\eta_1(x_0) \subset X$ and $\eta_1{}^*(x_0) \subset X^*$, where X^* is the conjugate space, is given by eq. (3.8).

In Section 4, the important case of bifurcation of solutions of eq. (2.1) at an exceptional point (x_0,λ_0) such that the null space $\eta_1(x_0)$ of $\lambda_0 I-T'(x_0)$ has dimension one, is taken up. Bifurcation eq. (3.8) is then one equation in one unknown, viz. eq. (4.1). The Newton Polygon method of solution is introduced and described, but for more details the reader is referred to the work of Dieudonné [ref. 8] and of Bartle [ref. 1 p. 376]. Next the eigenvalue problem $\lambda x = T(x)$ is specifically considered (i.e. we let $f = \theta$), as it is throughout the remainder of these notes.

The bifurcation equation (4.1) is specialized to the origin, i.e., we take $x_0 = \theta$, $\lambda_0 \neq 0$. At this point it is very convenient to assume that $T(x)$ is an odd operator, i.e., $T(-x) = -T(x)$ (statement H-2). With this added supposition, (θ,λ_0) being an exceptional point with respect to $T(x)$, there exist two nontrivial solution branches $x^{\pm}(\lambda)$ of the equation $T(x) = \lambda x$, consisting of eigenfunctions of $T(x)$, which bifurcate from the trivial solution $x = \theta$ at the point $\lambda = \lambda_0$. The two branches differ only in sign. The branching is to the left or right according to whether $a_2a_4 > 0$ or $a_2a_4 < 0$, where a_2, a_4 are coefficients in eq. (4.2), (see Theorem 4.2).

One of the purposes of section 5 is to explain in a general way the reasons for the subsequent preoccupation with Hammerstein operators, (see eq. (5.2)). Also two pure categories of nonlinearity are introduced for both Hammerstein operators and general completely continuous operators with symmetric Fréchet derivative, namely sublinearity and superlinearity; also the idea of asymptotic linearity is introduced. Bifurcation equation 4.2 and Theorem 4.2 are interpreted for Hammerstein operators. Lastly, the notion of an oscillation kernel is introduced together with the reason for its use: operator (5.6) has only simple eigenvalues whatever the continuous function $x(s)$ used in defining this operator.

In section 6 we take up the extension of branches of eigenfunctions of the Hammerstein operator (5.2) from the small into the large. Assumptions about the operator, mentioned in section 5, are given precise definition. Then it is shown that the branches of nontrivial solutions of the eigenvalue problem, given in the small by Theorem 4.2, which arise at the "primary" bifurcation points $\mu_n^{(o)}$, $n = 0,1,2,\cdots$, are such that the p'th branch functions have exactly p odd order interior zeros (and no other interior zeros). Actually this result holds in the large as well. Next, in the sublinear case it is shown that if, (a priori), $\Lambda_p < \frac{xf_x'(s,x)}{f(s,x)} < 1$, $0 \le s \le 1$, $-\infty < x < +\infty$, then $\lambda I - T'(x_p)$ cannot become singular so that there is no secondary bifurcation of the p'th branch. Here Λ_p is the number defined in connection with Lemma 6.4, and $T'(x_p)$ is the Fréchet derivative of the Hammerstein operator defined along the p'th branch $x_p = x_p^{\pm}(s,\lambda)$ of eigenfunctions. The analogous sufficient condition against secondary bifurcation in the superlinear case is $1 < \frac{xf_x'(s,x)}{f(s,x)} < \frac{1}{\Lambda_{p+1}}$, $0 \le s \le 1$, $-\infty < x < +\infty$. A couple of attempts are made to refine these conditions preventing secondary bifurcation of a branch of eigenfunctions by refining the definition of Λ_p. In the sublinear case, yet another condition against secondary bifurcation is given by inequality (6.15a), while a corresponding necessary condition is given by inequality (6.17); the two conditions bear some resemblance to each other but also show a conjugate feature, namely the occurrence of $\frac{f(s,\varphi)}{\varphi}$ in the sufficient condition, and $f_{x_p^*}'$ in the necessary condition.

Under Hypotheses H-1 through H-6, and under one or another of the conditions against secondary bifurcation, section 7 deals with branches of eigenfunctions of Hammerstein operators (5.2), in the large. Indeed, the p'th branch $x_p^{\pm}(s,\lambda)$ (the two parts differ only in sign) is a continuous locus of points in $C(0,1)$ which arises at the primary bifurcation point $\mu_p^{(o)}$, exists in the large, and becomes infinite as $\lambda \to \gamma_p$, where γ_p is the p'th eigenvalue of eq. (7.1). This again is under a condition against secondary bifurcation (such as 6.9 or 6.15) pertaining to the p'th branch. The number γ_p is sometimes called an "asymptotic bifurcation point."

Section 8 treats the example of section 1, previously solved by elementary methods, by the methods of functional analysis introduced in subsequent sections. For this example we know a precise condition preventing secondary bifurcation, namely $\frac{b}{a} \le \frac{1}{2}$ where a,b are the primary bifurcation points. It is shown that condition (6.9b) applied to this problem, and interpreted in terms of the primary bifurcation points, would give $\frac{b}{a} \le \frac{1}{3}$ as a condition at best; on the other hand condition (6.15b) actually yields $\frac{b}{a} \le \frac{1}{2}$. Hence, though (6.15b) is more complicated than condition (6.9b) and thus harder to apply, it is a more refined condition, at least for this example. Unfortunately we have no success in applying bifurcation equation (4.2) to study the secondary bifurcation of this example when it occurs, namely when $\frac{b}{a} > \frac{1}{2}$. Indeed, the leading coefficient of eq. (4.2) vanishes in this example at a secondary bifurcation point, but not enough of the lumped terms seem to cancel as was the case in producing eq. (4.4). Hence use of the Newton Polygon method as discussed by J. Dieudonné [ref. 8], and R. G. Bartle [ref. 1, p. 376] seems to founder on the requirement $\min_{1\le i\le n} \alpha_i = \min_{1\le i\le n} \beta_i = 0$ imposed upon the exponents of the bifurcation equation. Actually this failure is not to be expected in secondary bifurcation for Hammerstein operators which are such that $K(1-s,1-t) \ne K(s,t)$, or $f(1-s,x) \ne f(s,x)$.*

In section 9 we first note the equivalence between the eigenvalue problem $T(x) = \lambda x$ for the Hammerstein operator (5.2), where we assume

*Please see Appendix however.

that $K(s,t)$ has the form given in eq. (9.1), and a certain familiar two point boundary value problem, (9.2). It is known that in the autonomous case, $f(s,x(s)) \equiv f(x(s))$, there is no secondary bifurcation of <u>any</u> branch of solutions of problem (9.2). This is shown for a particular two-point boundary value problem, namely (9.4), in a way which clearly relates this absence of secondary bifurcation to some of our considerations of section 6. Eigenvalue problem (9.14) with kernel (9.15) generalizes the problem of section 1 in that the kernel is complete, (the kernel of section 1 had only the first two terms). But kernel (9.1) is that particular choice of kernel (9.15) in which we set $\mu_n^{(o)} = \frac{1}{n^2}$; also we note that the $\mu_n^{(o)}$'s are primary bifurcation points for problem (9.14). Hence by Theorem 9.3 there actually do exist sets of constants $\{\mu_n^{(o)}\}$ such that the eigenfunction branches arising at these primary bifurcation points undergo no secondary bifurcations at all. Indeed $\{\mu_n^{(o)}\} = \left\{\frac{1}{n^2}\right\}$ is one such set. Hence if we seek a condition on the primary bifurcation points $\{\mu_n^{(o)}\}$ of problem (9.14), with the complete kernel (9.15), such that there is no secondary bifurcation on <u>any</u> branch $\left(\frac{\mu_2^{(o)}}{\mu_1^{(o)}} \le \frac{1}{2}\right.$ is the condition if we use only the first two terms$\left.\right)$, the particular constants $\mu_n^{(o)} = \frac{1}{n^2}$ would presumably have to satisfy it.

Thus, assumptions H-1 through H-6 have been made on the nonlinear operator $T(x)$ mapping a real Banach space X into itself, with $T(\theta) = \theta$, and three times Fréchet differentiable. For the convenience of the reader we set down these cumulative hypotheses:

H-1: $T'(x_0)$ is a compact linear operator, where x_0 is a solution of the problem $T(x) = \lambda x$. This statement is fulfilled conveniently if the nonlinear operator $T(x)$ is completely continuous. Then $T'(x)$ is compact for any $x \in X$.

Statement H-1 is used in section 3.

H-2: $T(x)$ is an odd operator, i.e., $T(-x) = - T(x)$.

This is used in section 4. Further statements have to do with $T(x)$ as a Hammerstein operator in $C(0,1)$, (see eq. (5.2)). They are used in sections 6 and 7.

H-3: $f(s,x)$ is four times differentiable in x, with $|f_x^{iv}|$ bounded uniformly over $0 \le s \le 1$, and $\lim_{x\to 0} \frac{f(s,x)}{x} = f_x'(s,0)$ uniformly on $0 \le s \le 1$. (Statement H-2 already implies that $f(s,0) = 0$).

H-4a: Sublinearity; i.e. $f_x'(s,x) > 0$, $0 \le s \le 1$, and $xf_x''(s,x) < 0$, $0 \le s \le 1$, $-\infty < x < +\infty$.

H-4b: Superlinearity; i.e. $f_x'(s,x) > 0$, $0 \le s \le 1$, and $xf_x''(s,x) > 0$, $0 \le s \le 1$, $-\infty < x < +\infty$.

H-5: Asymptotic Linearity; i.e. $\lim_{|x|\to\infty} \frac{f(s,x)}{x} = A(s) \ge 0$ uniformly, $0 \le s \le 1$.

H-6: $K(s,t)$ is a symmetric oscillation kernel.

While these notes may answer some questions about nonlinear eigenvalue problems, it is probable that many more questions are raised.

Section 2 deals with a process for extending or continuing solutions of nonlinear equations in $X \times R$ under the assumption that we know a solution pair (x_0,λ_0) such that $\lambda_0 I - T'(x_0)$ has a bounded inverse. A more convenient continuation process might be based on solution of the vector differential equation $\frac{dx}{d\lambda} = -[\lambda I - T'(x)]^{-1}x$ relative to the initial condition $x = x_0$, $\lambda = \lambda_0$, provided one could carefully show the differential equation to be valid.

Almost immediately in section 3 we assumed that $T'(x_0)$ is compact; thus a discussion of other spectral alternatives--the continuous spectrum, the residual spectrum--was avoided. Without this assumption, points of $C\sigma T'$ and $R\sigma T'$ can be exceptional points with respect to the nonlinear transformation T. Thus one wonders, if (x_0,λ_0) is an exceptional point such that $\lambda_0 \in C\sigma T'(x_0)$ or $\lambda_0 \in R\sigma T'(x_0)$, whether or not eq. (2.3) ever leads to a bifurcation problem.

Again in section 3, though the derivation of the bifurcation equation (3.4) using the extended pseudo inverse has some aesthetic appeal, the arduous substitutions used to reduce this equation to manageable form, namely eq. (3.8), do not. The method used was that of Bartle [ref. 1, p. 370, p. 373]. M. S. Berger [ref. 4, pp. 127-136], N. W. Bazley and B. Zwahlen [ref. 23], and D. Sather [ref. 48] of the University of Wisconsin, seem to have methods which avoid these laborious substitutions. These authors make use of Krasnoselskii's bifurcation equation [ref. 14, pp. 229-231].*

In section 4 we considered only the case of bifurcation at an eigenvalue of multiplicity one. Bifurcation with higher multiplicity is a very difficult problem, though progress has been made by M. S. Berger [ref. 4, p. 134], Bazley and Zwahlen [ref. 3], Graves [ref. 12] and Sather [ref. 48]. This topic is important however since in studying secondary

*Please read the Appendix.

bifurcation, one cannot realistically assume that the multiplicity is one (which is why we have used the oscillation kernel in sections 5, 6 and 7). Again even if the secondary bifurcation multiplicity is one, there was trouble in applying the Newton Polygon method in the case where the leading coefficient a_1 in eq. (4.2) vanishes (see section 8). This problem needs attention in that this situation does arise with Hammerstein operators in which $K(1-s,1-t) = K(s,t)$ and $f(1-s,x) = f(s,x)$.

There are several unsettled questions arising from section 5. If there is no sublinearity or superlinearity condition, then we cannot prove Theorem 6.2. Accordingly we can get secondary bifurcation on the p'th branch in Hammerstein's equation through $\lambda = \mu_p$. These cases need study.* Likewise if we do not assume that $K(s,t)$ is an oscillation kernel, secondary bifurcation can arise through the possibility that on the p'th branch, $x_p^{\pm}(s,\lambda)$, we have $\lambda = \mu_n$ for $n > p$. Furthermore the secondary bifurcation could be with multiplicity greater than unity, as mentioned above. This provides a rich field for study. Again, is there a weaker assumption on the kernel $K(s,t)$ (other than assuming it to be an oscillation kernel) which would guarantee that the operator (5.6) has only simple eigenvalues, together with some identifiable way of enumerating the eigenfunctions (such as by the zeros), regardless of what continuous function $x(s)$ we use in defining the operator (5.6)? It was observed in section 5 that the oscillation kernel is actually in excess of our requirements as we know them now. Could such a weaker assumption about $K(s,t)$ enable us to study branches of eigenfunctions emanating from negative primary bifurcation points? (Oscillation kernels have only positive eigenvalues.) If we

*Also there is the case of nonsymmetric $K(s,t)$.

must live with the oscillation kernel, for the time being, in our desire to study branches of eigenfunctions in the large, how can this notion be generalized? Is there an abstract analogue, such as an "oscillation operator" in Hilbert space, which corresponds to the concrete idea of a linear integral operator in $L_2(0,1)$ with oscillation kernel? It is readily appreciated in these notes how our need for the oscillation kernel forces us into the study of Hammerstein operators rather than of a more abstract formulation.

With sections 6 and 7, the need is clearly that of a yet more refined condition against secondary bifurcation, both for the individual branches of eigenfunctions and for all the branches of eigenfunctions, under the given hypotheses. We need a sufficient condition which is also a necessary condition. If possible such conditions should be much more easily interpretable than either conditions (6.9) or (6.15). For example a condition might be devised on the primary bifurcation points $\{\mu_n^{(o)}\}$ which would preclude secondary bifurcation of a branch, or of any branch. In the example of section 1, we saw that $\frac{b}{a} > \frac{1}{2}$ was such a condition. Lastly, we seem to lack a good "figure of merit" for a sufficient condition against secondary bifurcation. How can one tell in a complicated problem which of several conditions is best without computation?

We have already mentioned the question which arises in section 8, that of the vanishing leading coefficient in the bifurcation equation (4.2) when we attempt to study the one secondary bifurcation of the problem; the case does not lend itself very well to the Newton Polygon analysis of Dieudonné or of Bartle. Can one modify the method, or find a new method?*

---

*Please see the Appendix.

How can one explain the lack of secondary bifurcation, on any branch of eigenfunctions in the autonomous case of boundary value problem (9.2), in terms of a condition against secondary bifurcation for the equivalent Hammerstein equation? What about the conditions of C. V. Coffman [ref. 6], and how do they relate to our conditions of section 6, if they relate at all? Again, the particular primary bifurcation points $\mu_n^{(o)} = \frac{1}{n^2}$ for eigenvalue problem (9.14), which is equivalent to boundary value problem (9.4), must satisfy <u>some</u> condition guaranteeing no secondary bifurcation of any eigenfunction branch. Theorem 9.3 proves there is no secondary bifurcation on the basis of the boundary value problem. What is this condition on the primary bifurcation points? Whatever the condition may be, we see that it is not vacuous.

There are many other questions in nonlinear analysis not arising specifically in these notes. Nonlinear eigenvalue problems occur in their own right in physics. We mention the problem of the buckled beam [ref. 14, pp. 181-183], the buckled circular plate [refs. 28, 35, 36, 53], the problem of waves on the surface of a heavy liquid [ref. 42], the problems of the heavy rotating chain and of the rotating rod [refs. 20 and 38], the problem of Taylor vortices in a rotating liquid [ref. 37], the Benard problem [refs. 27, 29, 35], and the problem of Ambipolar Diffusion of Ions and Electrons in a Plasma, [refs. 22, 24]. There are certainly many more.

Eigenfunctions for linear operators have useful properties apart from their immediate physical interpretation. One can prove such properties as completeness, which enables one to make expansions useful in solving inhomogeneous equations, the linear analogue of eq. (2.1). Are eigenfunctions for nonlinear operators ever complete in any sense, and is there a method of expansion in terms of these eigenfunctions which might be useful in connection with eq. (2.1)? The reader might wish to read some work by A. Inselberg on nonlinear superposition principles in this connection, [ref. 31].

It is well known that eigenfunctions for linear operators have much to do with the representation of solutions of initial value problems. Would eigenfunctions of nonlinear operators have anything to do with the representation of solutions of nonlinear initial value problems? In this connection, the reader might consult the recent work of Dorroh, Neuberger, Komura, and Kato on nonlinear semi-groups of nonexpansive transformations [refs. 25, 26, 32, 39, 43].

These notes have dealt largely with the point spectrum of the nonlinear operator $T(x)$. These are the points λ of the spectrum to which eigenfunctions are associated through the eigenvalue problem $T(x) = \lambda x$. Are there other kinds of points in the spectrum of $T(x)$? Under very broad assumptions J.Neuberger [ref. 44] has shown that a nonlinear operator does indeed have a spectrum. Under various assumptions on $T(x)$ is there a continuous spectrum or a residual spectrum, as with linear operators? Are there any other parts of the spectrum of these more general operators?

Workers in nonlinear Functional Analysis should not want for possible research paper topics.

BIBLIOGRAPHY

1. Bartle, R. G., "Singular points of functional equations," A.M.S. Transactions, Vol. 75 (1953), pp. 366-384.

2. Bazley, N. and B. Zwahlen, "Remarks on the bifurcation of solutions of a nonlinear eigenvalue problem," Archive for Rat. Mech. and Anal., Vol. 28 (1968), pp. 51-58.

3. Bazley, N. and B. Zwahlen, unpublished work by private communication.

4. Berger, M. S., "Bifurcation Theory for Real Solutions of Nonlinear Elliptic Partial Differential Equations," Lecture Notes on Bifurcation Theory and Nonlinear Eigenvalue Problems, edited by J. B. Keller and S. Antman, N.Y.U. Courant Inst. pp. 91-184.

5. Byrd, P. F., and M. D. Friedman, _Handbook of Elliptic Integrals for Engineers and Physicists_, Springer-Verlag, Berlin (1954).

6. Coffman, C. V., "On the uniqueness of solutions of a nonlinear boundary value problem," J. of Math. and Mech., Vol. 13 (1964) pp. 751-763.

7. Courant, R. and D. Hilbert, _Methods of Mathematical Physics, Vol. I_, Interscience Publishers, New York (1953).

8. Dieudonne, J., "Sur le Polygon de Newton," Archive der Math., Vol. 2 (1949-50), pp. 49-55.

9. Gantmacher, F. P. and M. G. Krein, _Oscillation matrices and kernels, and small vibrations of mechanical systems_, A.E.C. Transl. 4481; Office of Technical Services, Dept. of Commerce, Washington, D. C. (1961) $5. Available also in German Translation, Academic-Verlag, Berlin (1960).

10. Goldberg, S., _Unbounded Linear Operators_, McGraw Hill, 1966.

11. Graves, L., Theory of Functions of Real Variables, McGraw Hill, New York (1946).

12. Graves, L., "Remarks on singular points of functional equations," A.M.S. Transactions, Vol. 77 (1955), pp. 150-157.

13. Hildebrandt, T. H., and L. M. Graves, "Implicit functions and their differentials in general analysis," A.M.S. Transactions, Vol. 29 (1927) pp. 127-153.

14. Krasnoselskii, M. A., Topological methods in the theory of nonlinear integral equations, Pergamon Press, London (1964).

15. Lusternik, L. A., and V. I. Sobolev, Elements of functional analysis, Frederick Ungar Publishing Co., New York (1961).

16. Nirenberg, L., Functional Analysis Lecture Notes, N.Y.U. Courant Institute (1961).

17. Pimbley, G. H. Jr., "A sublinear Sturm-Liouville problem," J. of Math. and Mech., Vol. 11 (1962), pp. 121-138.

18. Pimbley, G. H. Jr., "The eigenvalue problem for sublinear Hammerstein operators with oscillation kernels," J. of Math. and Mech., Vol. 12 (1963) pp. 577-598.

19. Riesz, F. and B. Sz.-Nagy., Functional analysis, Frederick Ungar Publishing Co., New York (1955).

20. Tadjbakhsh, I., and F. Odeh, "A nonlinear eigenvalue problem for rotating rods," Archive for Rat. Mech. and Anal., Vol. 20 (1965) pp. 81-94.

21. Zaanen, A. C., Linear analysis, Interscience Publishers Inc., New York (1953).

ADDITIONAL REFERENCES

22. Allis, W. P. and D. J. Rose, "The transition from free to ambipolar diffusion," The Physical Review, Vol. 93 (1954), No. 1, pp. 84-93.

23. Bazley, N. W. and B. Zwahlen, "Estimation of the bifurcation coefficient for nonlinear eigenvalue problems," Battelle Institute Advanced Studies Center (Geneva, Switz.), Math. Report No. 16, Oct. 1968.

24. Dreicer, H., "Ambi-polar diffusion of ions and electrons in a magnetic field," Interoffice communication at Los Alamos Scientific Laboratory (1955).

25. Dorroh, J. R., "Some classes of semi-groups of nonlinear transformations and their generators," J. of Math. Soc. of Japan, Vol. 20 (1968), No. 3, pp. 437-455.

26. Dorroh, J. R., "A nonlinear Hille-Yosida-Phillips theorem," to appear in Journal of Functional Analysis, Vol. 3 (1969).

27. Fife, P. C. and D. D. Joseph, "Existence of convective solutions of the generalized Bénard problem which are analytic in their norm," to appear in Archives for Rational Mechanics and Analysis.

28. Friedrichs, K. O. and J. J. Stoker, "The non-linear boundary value problem of the buckled plate," Amer. Jour. of Math., Vol. 63 (1941), pp. 839-888.

29. Görtler, H., K. Kirchgässner and P. Sorger, "Branching solutions of the Bénard problem," to appear in Problems of Hydrodynamics and Continuum Mechanics, dedicated to L. I. Sedov.

30. Hammerstein, A., "Nichtlineare integralgleichungen nebst anwendungen," Acta Mathematica, Vol. 54 (1930), pp. 117-176.

31. Inselberg, A., "On classification and superposition principles for nonlinear operators," A. F. Grant 7-64 Technical Report No. 4, May 1965, sponsored by the Air Force Office of Scientific Research, Washington 25, D. C.

32. Kato, T., "Nonlinear semigroups and evolution equations," J. Math. Soc. of Japan, Vol. 19 (1967), pp. 508-520.

33. Keller, H. B., "Nonlinear bifurcation," by private communication from H. B. Keller, Calif. Inst. of Tech., 1969.

34. Keller, H. B., "Positive Solutions of some Nonlinear Eigenvalue Problems," to appear in J. Math. and Mech., Oct., 1969.

35. Keller, J. B. and S. Antman, "Bifurcation theory and nonlinear Eigenvalue Problems," Lecture notes, Courant Institute of Mathematical Sciences, New York University, 1968.

36. Keller, J. B., Keller, H. B. and E. L. Reiss, "Buckled states of circular plates," Q. Appl. Math., Vol. 20 (1962), pp. 55-65.

37. Kirchgässner, K. and P. Sorger, "Stability analysis of branching solutions of the Navier-Stokes equations," by private communication from K. Kirchgässner, Freiburg University, W. Germany.

38. Kolodner, I. I., "Heavy rotating string--a nonlinear eigenvalue problem," Comm. on Pure and Applied Math., Vol. 8 (1955), pp. 395-408.

39. Komura, Y., "Nonlinear semi-groups in Hilbert space," J. Math. Soc. of Japan, Vol. 19 (1967), pp. 493-507.

40. Krasnoselskii, M. A., Positive Solutions of Operator Equations, Chap. 6, section 6.4, P. Noordhoff Ltd., Groningen, Netherlands, 1964.

41. Krasnoselskii, M. A. "Some problems of nonlinear analysis," A.M.S. Translations, series 2, Vol. 10 (1958), pp. 345-409.

42. Nekrasov, A. I., "The exact theory of steady waves on the surface of a heavy fluid," translated as a U. of Wisconsin Report: MRC-TSR 813.

43. Neuberger, J. W., "An exponential formula for one-parameter semi-groups of nonlinear transformations," J. Math. Soc. of Japan, Vol. 18 (1966), pp. 154-157.

44. Neuberger, J. W., "Existence of a spectrum for nonlinear transformations," to appear in Pacific Journal of Mathematics.

45. Pimbley, G. H. Jr., "A fixed-point method for eigenfunctions of sub-linear Hammerstein operators," Archive for Rat. Mech. and Anal., Vol. 31 (1968).

46. Pimbley, G. H. Jr., "Positive solutions of a quadratic integral equation," Archive for Rat. Mech. and Anal., Vol. 24 (1967), No. 2, pp. 107-127.

47. Pimbley, G. H. Jr., "Two conditions preventing secondary bifurcations for eigenfunctions of sublinear Hammerstein operators," to appear in J. of Math. and Mech., 1969.

48. Sather, D., "Branching of solutions of an equation in Hilbert space," to appear as a Mathematics Research Center Report, U. S. Army, University of Wisconsin.

49. Tchebotarev, N. G., "Newton's polygon and its role in the present development of mathematics," Papers presented on the occasion of the tri-centenary of Isaac Newton's birth, Moscow, 1943, pp. 99-126. Copy available in U. S. Library of Congress. Translation at Los Alamos Scientific Laboratory library, 512C514mt.

50. Vainberg, M. M. and P. G. Aizengendler, "The theory and methods of investigation of branch points of solutions," Progress in Mathematics, Vol. 2, edited by R. V. Gamkrelidge, Plenum Press, New York, 1968.

51. Vainberg, M. M. and V. A. Trenogin, "The methods of Lyapunov and Schmidt in the theory of nonlinear equations and their further development, "Uspekhi Matematicheskikh Nauk (English transl.), Vol. 17 (1962), No. 2, pp. 1-60.

52. Wing, G. M., and L. F. Shampine, "Existence and uniqueness of solutions of a class of nonlinear elliptic boundary value problems," by private communication from G. M. Wing, University of New Mexico, Albuquerque, N. M.

53. Wolkowisky, J. H., "Existence of buckled states of circular plates," Comm. on Pure and Applied Math., Vol. XX (1967), pp. 549-560.

APPENDIX: ANOTHER BIFURCATION METHOD, THE EXAMPLE OF SECTION 1, RECONSIDERED AGAIN.

In Section 3 we presented quite a traditional method of converting the bifurcation equation, eq. (3.4) into a form suitable for application of the Newton Polygon method of solution. It involved substitution of the nonlinear expansion, eq. (3.5), into expression (3.7) so as to compute the quantity $F_\delta(V_\delta(u))$. As the patient reader discovers, this method is very arduous.

Several recent authors, notably L. Graves [ref. 12], M. S. Berger [ref. 4, p. 37], N. Bazley and B. Zwahlen [ref. 23], and D. Sather [ref. 48], succeed in discussing bifurcation without this arduous substitution process. Their treatments are rather special, however, in that they apply only in the case of bifurcation at the origin ($x_0 = \theta$ in Section 3). In this Appendix we present a treatment, patterned after those mentioned above, with which bifurcation can be treated anywhere (i.e., $x_0 \neq \theta$). For simplicity, however, we confine the treatment here to the case of unit multiplicity (i.e., n = 1); the above authors treat cases of higher multiplicity as well.

Let us digress at the outset to present a way of reducing eq. (2.3) (where $\lambda_0 I - T'(x_0)$ is singular) to eq. (3.2) and eq. (3.4), which is more simple than that appearing in Section 3. It is correct in the case where λ_0 is an eigenvalue of the compact linear operator $T'(x_0)$ of Riesz index unity, [ref. 21, p. 334, Th. 6]. In particular it works then in the case when $T'(x_0)$ is a compact, symmetric or symmetrizable operator on a real Hilbert space H, [ref. 21, p. 342, Theorem 18, p. 397, Theorem 3]. This reduction is usually attributed to M. A. Krasnoselskii, [ref. 14, p. 229].

If λ_o has Riesz index unity, then by definition, $X = \eta_1(x_o) \oplus R_1(x_o)$, and there exist complementary projections E_1, E^1 which project respectively on $\eta_1(x_o)$ and $R_1(x_o)$, and which commute with $T'(x_o)$. Moreover, $\lambda_o I - T'(x_o)$ has a bounded inverse on the subspace $R_1(x_o) = E^1 X$.

We premultiply eq. (2.3) by the projection E^1 to obtain

$$[\lambda_o I - T'(x_o)]E^1 h = -\delta E^1 x_o + \delta E^1 h + E^1 R_1(x_o, h)$$

$$= E^1 F_\delta(h).$$

Letting $h = k + u$, where $k \in R_1(x_o)$, $u \in \eta_1(x_o)$, gives

$$k = [\lambda_o I - T'(x_o)]^{-1} E^1 F_\delta(k+u). \tag{A.1}$$

Now $[\lambda_o I - T'(x_o)]^{-1} E^1$ is just the extended pseudo-inverse discussed in Section 3. Adding u to both sides of eq. (A.1) results in eq. (3.2), to which Theorem 3.2 can be applied to produce the local solution $h = V_\delta(u)$.

Again if we premultiply eq. (2.3) by the projection E_1, we have

$$E_1 F_\delta(h) = (I - E^1)F_\delta(h) = [\lambda_o I - T'(x_o)]E_1 h$$

$$= \theta$$

which is just eq. (3.4).

Thus when the index of λ_o, as an eigenvalue of the compact operator $T'(x_o)$, is unity, the approach to the bifurcation equations (3.2) and (3.4) is straightforward.

Let us now return to the bifurcation analysis based on equation (3.2) and (3.4). When the <u>multiplicity</u> of λ_o is also unity, we set $u = \xi u_o$ with ξ a real scalar and u_o the normalized element spanning $\eta_1(x_o)$. In the sequel it will be convenient to redesignate the local solution of eq. (3.2), given by Theorem 3.2, as $h = V(\delta,\xi)$. We also generalize the expansion (3.7) used in Section 3:

$$F_\delta(h) = -\delta x_o - \delta h + C_{x_o}(h) + D_{x_o}(h), \tag{A.2}$$

where $\alpha! C_{x_o}(h)$ is the first nonvanishing symmetric Fréchet differential form [ref. 15, p. 188] at x_o, homogeneous of integer order $\alpha \geq 2$, $\left(\text{viz. } C_{x_o}(h) = \frac{1}{\alpha!} d^\alpha T(x_o; \overbrace{h,\cdots,h}^{\alpha})\right)$, and $D_{x_o}(h) = R_\alpha(x_o;h)$ is the remainder, with $\lim_{\|h\|\to 0} \frac{\|D_{x_o}(h)\|}{\|h\|^\alpha} = 0$.

Recalling now the bifurcation equation in the form (3.6), where $u_o^* \in X^*$ is the normalized eigenelement of the adjoint operator $T'(x_o)^*$, we write eq. (3.4) as

$$-\delta u_o^* x_o - \delta\xi u_o^* u_o + u_o^* C_{x_o}(V(\delta,\xi)) + u_o^* D_{x_o}(V(\delta,\xi)) = 0. \tag{A.3}$$

We now seek to pass directly from eq. (A.3) to a form which resembles eq. (4.3) and to which the Newton polygon type of analysis may be applied. Namely, from eq. (A.3) we write:

$$\begin{aligned} &\delta u_o^* x_o + \delta\xi u_o^* u_o \\ &\quad = u_o^*[C_{x_o}(\xi u_o) + C_{x_o}(V(\delta,\xi)) - C_{x_o}(\xi u_o)] + u_o^* D_{x_o}(V(\delta,\xi)) \qquad (A.4) \\ &\quad = \xi^\alpha u_o^* C_{x_o}(u_o)\left[1 + \frac{u_o^*\left\{C_{x_o}(V(\delta,\xi)) - C_{x_o}(\xi u_o)\right\}}{\xi^\alpha u_o^* C_{x_o}(u_o)} + \frac{u_o^* D_{x_o}(h)}{\xi^\alpha u_o^* C_{x_o}(u_o)}\right]. \end{aligned}$$

Following Bazley and Zwahlen [ref. 23, p. 4], we must show that the quantity in brackets in eq. (A.4) tends to unity as $\delta,\xi \to 0$. To this end we have the following lemmas:

Lemma A.1: $V_\xi(\delta,\xi)$ exists and is continuous in its arguments; i.e. the solution $h = V(\delta,\xi)$ of eq. (3.2), (with $u = \xi u_0$), is continuously differentiable with respect to ξ.

Proof: The result follows from the generalized Implicit Function Theorem (which yields the same result about eq. (3.2) as our Theorem (3.2), if not the same nonlinear development (3.5) needed in Section 3) and its corollary about derivatives [ref. 15, p. 194, Th. 1; p. 196, Th. 2].

Lemma A.2: $\lim\limits_{\substack{\xi\to 0\\ \delta\to 0}} \frac{V(\delta,\xi)}{\xi} = u_0.$

Proof: From eq. (3.2) we form the following difference equation:

$$\frac{V(\delta,\xi+t)-V(\delta,\xi)}{t} = \overline{M}(x_0)\ \frac{F_\delta(V(\delta,\xi+t)-F_\delta(V(\delta,\xi))}{t} + u_0.$$

Using eq. (A.2) and passing to the limit as $t \to 0$, we get the following equation for $V_\xi(\delta,\xi)$:

$$V_\xi = \overline{M}(x_0)\Big\{ -\delta V_\xi + dC_{x_0}(V(\delta,\xi);V_\xi) + dD_{x_0}(V(\delta,\xi);V_\xi)\Big\} + u_0; \qquad \text{(A.5)}$$

here $dC_{x_0}(h;k)$ and $dD_{x_0}(h;k)$ are the first Fréchet differentials of the nonlinear operators $C_{x_0}(h)$, $D_{x_0}(h)$ at $h_0 \in X$.

It can be verified that $C_{x_0}(h) = \frac{1}{\alpha!}\, d^\alpha T(x_0;\overbrace{h,\cdots,h}^{\alpha})$ has a first Fréchet differential $dC_{x_0}(h,k)$ which satisfies the inequality

$$\|dC_{x_0}(h,k)\| \le \frac{1}{(\alpha-1)!}\, K(x_0)\|h\|^{\alpha-1}\cdot\|k\|$$

where $K(x_0) > 0$ is a constant for given $x_0 \in X$.

Again, $D_{x_0}(h) = T(x_0+h) - T(x_0) - T'(x_0)h - C_{x_0}(h)$ is continuously Fréchet differentiable, so that $\lim_{h\to\theta} dD_{x_0}(h;k) = dD_{x_0}(\theta;k)$, uniformly for $\|k\| = 1$. Moreover since $D_{x_0}(\theta) = \theta$, we have, using the definition of the Fréchet differential [ref. 15, p. 183],

$$dD_{x_0}(\theta;k) = \lim_{\sigma\to 0}\left[\frac{D_{x_0}(\sigma k)}{\sigma} - \frac{E_{x_0}(\theta;\sigma k)}{\sigma}\right] = \theta$$

since $E_{x_0}(\theta;k) = o(\|k\|)$ and $D_{x_0}(h) = o(\|h\|^{\alpha})$. Thus $\lim_{\|h\|\to 0} dD_{x_0}(h;k) = \theta$ uniformly on $\|k\| = 1$.

We now recall an estimate for $V(\delta,\xi)$ derived in Section 3 and appearing at the top of page 26; namely

$$\|V(\delta,\xi)\| \le 2(\|\overline{M}\|\cdot\|x_0\|\,|\delta| + \|u_0\|\cdot|\xi|),\ \|h\| \le d_2,\ |\delta| \le \delta_2, \tag{A.6}$$

where d_2, δ_2 are numbers arising in the proof of Theorem 3.2. Thus $\|V(\delta,\xi)\| \to 0$ as $\delta,\xi \to 0$. Using eq. (A.5) we then have, for sufficiently small δ,ξ, the estimate:

$$\|V_\xi(\delta,\xi)\| \le \frac{2\|u_0\|}{1-2\|\overline{M}\|\{\alpha K(x_0)\|V(\delta,\xi)\|^{\alpha-1}+\|D'_{x_0}(V(\delta,\xi))\|}$$

where $\|D'_{x_0}(h)\| = \sup_{\|k\|=1}\|dD_{x_0}(h;k)\|$, and where, as was shown above, $\lim_{\|h\|\to 0}\|D'_{x_0}(h)\| = 0$. This shows that $\|V_\xi(\delta,\xi)\|$ remains bounded as $\delta,\xi \to 0$; say $\|V_\xi\| \le M_1$.

Using eq. (A.5) a second time, we have

$$\|V_\xi(\delta,\xi)-u_0\| \le \|\overline{M}\|\{|\delta|M_1+\alpha K(x_0)\cdot\|V(\delta,\xi)\|^{\alpha-1}M_1 + \|D'_{x_0}(V(\delta,\xi))\|M_1\},$$

so that again by virtue of the estimate (A.6) above, we see that $\lim\limits_{\substack{\xi\to 0\\ \delta\to 0}} \frac{V(\delta,\xi)}{\xi} = \lim\limits_{\substack{\xi\to 0\\ \delta\to 0}} V_{\xi}(\delta,\xi) = u_o$. This proves the lemma.

We return now to expression (A.4) for the bifurcation equation. We note the following estimates:

$$\left| \frac{u_o^*\left\{C_{x_o}(V(\delta,\xi)) - C_{x_o}(\xi u_o)\right\}}{\xi^{\alpha} u_o^* C_{x_o}(u_o)} \right|$$

$$\leq \frac{\left\| C_{x_o}\left(\frac{V(\delta,\xi)}{\xi}\right) - C_{x_o}(u_o) \right\|}{|u_o^* C_{x_o}(u_o)|} \to 0$$

as $|\delta|, |\xi| \to 0$

in view of the α-homogeneity of $C_{x_o}(h)$, and Lemma A.2. Also

$$\left| \frac{u_o^* D_{x_o}(V(\delta,\xi))}{\xi^{\alpha} u_o^* C_{x_o}(u_o)} \right| \leq \frac{\| D_{x_o}(V(\delta,\xi))\|}{\|V(\delta,\xi)\|^{\alpha}} \cdot \left| \frac{\|V(\delta,\xi)\|}{\xi} \right|^{\alpha} \cdot \frac{1}{|u_o^* C_{x_o}(u_o)|}$$

$\to 0$ as $\delta,\xi \to 0$

by virtue of the property $D_{x_o}(h) = o(\|h\|^{\alpha})$ and estimate (A.6).

We now see, with reference to eq. (A.4), that the bifurcation equation for the case $n = 1$ can be written

$$\delta u_o^* x_o + \delta\xi u_o^* u_o = \xi^{\alpha} u_o^* C_{x_o}(u_o)[1 + \Phi(\delta,\xi)] \qquad (A.7)$$

where

$$\Phi(\delta,\xi) = \frac{u_o^*\left\{C_{x_o}(V(\delta,\xi)) - C_{x_o}(\xi u_o)\right\} + u_o^* D_{x_o}(V(\delta,\xi))}{\xi^{\alpha} u_o^* C_{x_o}(u_o)}$$

tends to zero as $\delta,\xi \to 0$.

Thus we have a bifurcation equation of the form (4.3) which can be treated by the Newton Polygon method. In a manner similar to that in Section 4, we can label the coefficients:

$$a_1 = -u_o{*}x_o, \quad a_2 = -u_o{*}u_o,$$

$$a_{\alpha+1} = u_o{*}C_{x_o}(u_o) = \frac{1}{\alpha!}\, u_o{*}d^{\alpha}T(x_o;\overbrace{u_o,\cdots,u_o}^{\alpha}),$$

integer $\alpha \geq 2$.

Bifurcation at the origin, $x_o = 0$, leads to $a_1 = 0$. We then have Theorems 4.2 and 4.3, proved in exactly the same way as in Section 4. $\Phi(\delta,\xi)$ can be shown to have the needed differentiability. (Let us note here that $a_2 \neq 0$ needs justification unless X is a Hilbert space. Likewise $C_{x_o}(h)$ might more properly be taken as the first Fréchet differential form such that $a_{\alpha+1} \neq 0$.)

It is interesting to see how we can resolve a difficulty which arose in Section 8 using the form of the bifurcation equation, namely (A.7), derived in this Appendix.

For Hammerstein operators with symmetric kernel, we have derived specific coefficients for the bifurcation equation, namely those designated eq. (5.3). The corresponding coefficients for eq. (A.7) are

$$a_1 = -(f'_{x_o}u_o,x_o)\ , \quad a_2 = -(f'_{x_o}u_o,u_o), \quad a_{\alpha+1} = \frac{1}{\alpha!}\lambda_o\,(f^{(\alpha)}_{x_o},u_o^{\alpha+1}),$$

integer $\alpha \geq 2$.

For the example being discussed in Sections 1 and 8, these coefficients become:

$$a_1 = -([1+3\varphi_{sb}^2]\sin 2s, \varphi_{sb}) = 0,$$

$$a_2 = -([1+3\varphi_{sb}^2]\sin 2s, \sin 2s)$$

$$= -\int_0^\pi \sin^2 2s ds - 4\left(\frac{a-b}{2b-a}\right)\int_0^\pi \sin^2 s \sin^2 2s ds$$

$$= -\left[1 + \frac{a-b}{2b-a}\right]\pi < 0,$$

$$a_3 = \frac{1}{2!}\frac{ab}{2b-a}(6\varphi_{sb}, \sin^3 2s)$$

$$= \pm 3 \frac{ab}{2b-a}\int_0^\pi \frac{2}{\sqrt{3}}\sqrt{\frac{a-b}{2b-a}} \sin s \sin^3 2s ds = 0$$

$$a_4 = \frac{ab}{2b-a}(1, \sin^4 2s) = \frac{ab}{2b-a}\int_0^\pi \sin^4 2s ds$$

$$= \frac{3}{8}\pi \frac{ab}{2b-a} > 0,$$

where we recall from Section 8 that $\varphi_{sb} = \pm \frac{2}{\sqrt{3}}\sqrt{\frac{a-b}{2b-a}} \sin s$ are the points on the zero'th branch at $\lambda = \lambda_{sb} = \frac{ab}{2b-a}$ where secondary bifurcation takes place.

With these coefficients, eq. (A.7) becomes

$$a_2\delta\xi + a_4\xi^3[1+\Phi(\delta,\xi)] = 0. \qquad (A.8)$$

Eq. (A.8) is handled in the same way as similar equation (4.4) is treated in the proof of Theorem 4.2. Since $a_2a_4 < 0$, bifurcation is to the right, as is certainly indicated in Section 1. There is a trivial solution $\xi = 0$

of eq. (A.8) which when substituted in eq. (3.2), along with $x_o = \varphi_{sb}$, leads to the continuation of a solution proportional to sin s, again as expected from Section 1. Two other nontrivial solutions of eq. (A.8), namely $\xi^{\pm}(\delta)$, lead, through eq. (3.2), to a pair of solutions which branch away from the secondary bifurcation point $(\varphi_{sb},\lambda_{sb})$. Again these branching solutions can be observed by means of eq. (3.2), which we can now write as

$$h = M\left\{-\delta E^1\varphi_{sb} - \delta E^1 h + \frac{2}{\pi}\int_0^{\pi} a \sin s \sin t \left[3\varphi_{sb}(t)h^2(t)+h^3(t)\right]dt\right\}$$

$$+ \xi^{\pm}(\delta)\sin 2s,$$

to be expressed as linear combinations of sin s and sin 2s. (E^1 is the projection on the range spanned by sin s.)

In the study of secondary bifurcation, at least with the example of Section 1, the traditional bifurcation method as set forth in Section 3 seems to result in unwanted terms which are troublesome when a_1 vanishes. The method described in this Appendix seems to be free of such terms. We have yet to obtain a good understanding of this, and to reconcile the two methods in this respect.

Offsetdruck: Julius Beltz, Weinheim/Bergstr

Bitte wenden / Continued

Beschaffenheit der Manuskripte

Die Manuskripte werden photomechanisch vervielfältigt; sie müssen daher in sauberer Schreibmaschinenschrift geschrieben sein. Handschriftliche Formeln bitte nur mit schwarzer Tusche eintragen. Notwendige Korrekturen sind bei dem bereits geschriebenen Text entweder durch Überkleben des alten Textes vorzunehmen oder aber müssen die zu korrigierenden Stellen mit weißem Korrekturlack abgedeckt werden. Falls das Manuskript oder Teile desselben neu geschrieben werden müssen, ist der Verlag bereit, dem Autor bei Erscheinen seines Bandes einen angemessenen Betrag zu zahlen. Die Autoren erhalten 75 Freiexemplare.

Zur Erreichung eines möglichst optimalen Reproduktionsergebnisses ist es erwünscht, daß bei der vorgesehenen Verkleinerung der Manuskripte der Text auf einer Seite in der Breite möglichst 18 cm und in der Höhe 26,5 cm nicht überschreitet. Entsprechende Satzspiegelvordrucke werden vom Verlag gern auf Anforderung zur Verfügung gestellt.

Manuskripte, in englischer, deutscher oder französischer Sprache abgefaßt, nimmt Prof. Dr. A. Dold, Mathematisches Institut der Universität Heidelberg, Tiergartenstraße oder Prof. Dr. B. Eckmann, Eidgenössische Technische Hochschule, Zürich, entgegen.

Cette série a pour but de donner des informations rapides, de niveau élevé, sur des développements récents en mathématiques, aussi bien dans la recherche que dans l'enseignement supérieur. On prévoit de publier

1. des versions préliminaires de travaux originaux et de monographies
2. des cours spéciaux portant sur un domaine nouveau ou sur des aspects nouveaux de domaines classiques
3. des rapports de séminaires
4. des conférences faites à des congrès ou à des colloquiums

En outre il est prévu de publier dans cette série, si la demande le justifie, des rapports de séminaires et des cours multicopiés ailleurs mais déjà épuisés.

Dans l'intérêt d'une diffusion rapide, les contributions auront souvent un caractère provisoire; le cas échéant, les démonstrations ne seront données que dans les grandes lignes. Les travaux présentés pourront également paraître ailleurs. Une réserve suffisante d'exemplaires sera toujours disponible. En permettant aux personnes intéressées d'être informées plus rapidement, les éditeurs Springer espèrent, par cette série de »prépublications«, rendre d'appréciables services aux instituts de mathématiques. Les annonces dans les revues spécialisées, les inscriptions aux catalogues et les copyrights rendront plus facile aux bibliothèques la tâche de réunir une documentation complète.

Présentation des manuscrits

Les manuscrits, étant reproduits par procédé photomécanique, doivent être soigneusement dactylographiés. Il est recommandé d'écrire à l'encre de Chine noire les formules non dactylographiées. Les corrections nécessaires doivent être effectuées soit par collage du nouveau texte sur l'ancien soit en recouvrant les endroits à corriger par du verni correcteur blanc.

S'il s'avère nécessaire d'écrire de nouveau le manuscrit, soit complètement, soit en partie, la maison d'édition se déclare prête à verser à l'auteur, lors de la parution du volume, le montant des frais correspondants. Les auteurs recoivent 75 exemplaires gratuits.

Pour obtenir une reproduction optimale il est désirable que le texte dactylographié sur une page ne dépasse pas 26,5 cm en hauteur et 18 cm en largeur. Sur demande la maison d'édition met à la disposition des auteurs du papier spécialement préparé.

Les manuscrits en anglais, allemand ou francais peuvent être adressés au Prof. Dr. A. Dold, Mathematisches Institut der Universität Heidelberg, Tiergartenstraße ou au Prof. Dr. B. Eckmann, Eidgenössische Technische Hochschule, Zürich.